Márcio Rodrigues Silva

Unveiling the city

Márcio Rodrigues Silva

Unveiling the city

Socio-spatial segregation in Jataí

ScienciaScripts

Imprint
Any brand names and product names mentioned in this book are subject to trademark, brand or patent protection and are trademarks or registered trademarks of their respective holders. The use of brand names, product names, common names, trade names, product descriptions etc. even without a particular marking in this work is in no way to be construed to mean that such names may be regarded as unrestricted in respect of trademark and brand protection legislation and could thus be used by anyone.

Cover image: www.ingimage.com

This book is a translation from the original published under ISBN 978-3-330-76412-5.

Publisher:
Sciencia Scripts
is a trademark of
Dodo Books Indian Ocean Ltd. and OmniScriptum S.R.L publishing group

120 High Road, East Finchley, London, N2 9ED, United Kingdom
Str. Armeneasca 28/1, office 1, Chisinau MD-2012, Republic of Moldova, Europe
Managing Directors: Ieva Konstantinova, Victoria Ursu
info@omniscriptum.com

Printed at: see last page
ISBN: 978-620-8-41184-8

To Sinara, my dear wife and best friend. To little Julia and Luisa.

ACKNOWLEDGEMENTS

To God.

To my dear wife Sinara, my companion at all times.

To my parents and sisters, for their solid family foundation. Examples of love, struggle and perseverance.

To my colleagues on the Geography course at the Jataí Campus - UFG.

To the master, Prof Dr Iraci Scopel, for his ever-present encouragement.

To Prof Dr João Batista de Deus, for believing in my work and for his serious guidance.

Sola Gratia

Sola Fide

Sola Scriptura

Solus Christus

Soli Deo Gloria

SUMMARY

Since analysing urban problems in order to chart new directions for local development is fundamental to improving the quality of life of the population, the focus of this research is to study the urban structure of Jataí-GO. To this end, its internal structure is analysed in terms of socio-spatial distribution in the urban arrangement, seeking answers to the question: is there socio-spatial segregation in Jataí? The aim is to unveil the city and propose new parameters that contribute to local development, as well as to produce consistent scientific material on this issue analysed according to Geographical Science. Jataí is located in south-west Goiás, approximately 320 kilometres from the capital, Goiânia. With an urban area of approximately 26km^2, within a municipality with an area of 7,174.10km^2, the city has expanded marked by contrasts that reveal the logic of capitalist society. Positions such as those of Santos (1993), Castells (1983), Lojkine (1997) and Villaça (2001), raising the logic of the formation and structuring of urban space, raise questions about the object of study in question. The path followed in this work is not rigid, ready and finished; it is an analysis that starts from the urban landscape, since this presents a vast horizon to be explored, arriving at the place and analysing how it was produced. The research moves through political and social analyses, contextualised in a so-called "post-modern" society, questioning and reflecting on the reality presented before our eyes, considering the action of capital and its relations in the construction of places. In this way, the dialectical perspective is presented as a guiding path for the work. In the end, we defined, classified, spatialised and cross-referenced information plans relating to income levels in Jataí and their implications for the urban fabric, demonstrating the urban segregation identified. The results are presented in the form of text and maps that reveal the real situation of the socio-spatial distribution in question. The results show that the lower income classes are predominantly concentrated on the edges of the urban area. This analysis is in line with the thesis that this is a segregated space.

Key words: Urban Segregation; Urban Structure; City;

Jataí-GO.

CONTENTS

INITIAL CONSIDERATIONS

This work is a continuation of the questions that arose during the course of learning geography as an undergraduate. Positions such as those of Santos (1993), Castells (1983), Lojkine (1997) and Villaça (2001), raising the logic of the formation and structuring of urban space, raise questions about the object of study of this work: the city of Jataí-GO. Located in the south-west of the state of Goiás, Jataí originated in the first half of the 19th century as a result of the search for new areas to develop agriculture and currently occupies an important place in the state's economy, being a leading producer of grains, which began to occur mainly in the 1970s, with the arrival and subsequent consolidation of modern agriculture in the municipality.

A structured space that doesn't arise at random, with a range of variables that make it unique. These are characteristics that are pertinent to urban centres, whether large or small, and which have aroused a researcher's gaze towards the city of Jataí.

With an urban area of approximately 26km^2, within a municipality with an area of 7,174.10km^2, Jataí has expanded marked by contrasts that reveal the logic of capitalist society. It is a contradictory space, a reflection of a society based on principles that distance themselves from valuing the individual as a citizen, considering them merely as a number in statistical data. A contradictory space where the "new" is produced to the detriment of the traditional, which becomes obsolete in the process of reproducing urban space. It's a fragmented space, with places not previously imagined , where social relations are restricted to the bare minimum and social experiences are reduced to the private sphere; perhaps because of the lack of time in modern society, perhaps because of the lack of identity with the "new" in a city that retains a lot of rural relations in its traditions.

In this sense, as Seabra (2001) emphasises, the role of the geographer is to understand how life is being represented, how socio-spatial relations take place within a dialectical vision.

Thus, insofar as the socio-spatial distribution of the population in the urban area expresses contradictions that are not perceived at first glance, the aim was to analyse the internal structuring of Jataí, with regard to socio-spatial distribution in the urban arrangement, seeking answers to the question: is there socio-spatial segregation in Jataí?

The work is centred mainly on the post-1970s scenario, with the aim of unveiling the city and proposing new parameters that collaborate with local development. To this end, economic, social and political data is analysed and spatialised to reveal the city's production process. Always bearing in mind that the object of study is part of a larger context and is involved in the reality of the state, the country and the world, suffering influences and influencing these spheres.

In order to achieve the objectives, the studies started from the path previously traced in the master's research entitled "Encontros e Desencontros: Estudo do Espaço Urbano de Jataí-GO" (Encounters and Disencounters: A Study of the Urban Space of Jataí-GO) (SILVA, 2005). Taking into account the scant specific bibliography on the urban study of Jataí, this being the second specific work on the urban area, we sought to build a consistent and well-founded base on the subject, as a way of supporting future work on this important city in south-west Goiás.

As part of the work, I was given the task of constructing the entire cartographic base and the analysis maps relating to the urban space of Jataí used in this research, a task that starts the proposed analysis at a time of reflection on the city through cartographic constructions.

Most of the data specific to the city required enquiries at the relevant organisations, and was basically found at the local town hall and the IBGE.

In this context, the results obtained from the research consolidate the ballast needed for discussions about Jataí's urban scenario.

The following research sources were fundamental to understanding Jatai's urban space: The Municipal Finance Department's property register for 2003 - where data was collected on urban voids and property values; the Fiscal Management Department's 2004 programme for registering economic activities - which contributed economic data; a report on economic activities registered up to 2004, also from the Fiscal Management Department - making it possible to see the evolution of economic activities set up in the city; Urban Master Plan of 2001, 2003, 2005, 2007 and 2008 - allowing the municipality's urban policies to be verified, preliminary survey of vacant plots in 1995 - which served as a parameter for a comparative analysis of urban voids, cadastral maps of 2003 and 2004 - provided on digital media, which greatly facilitated the preparation of the cartographic material, all provided by the Urban Planning Division of the Department of Works and Urbanism; IBGE Information Base by Census Sectors for the year 2000 - fundamental material, with information from the 2000 census for the city's urban area, which provided essential income data for discussions on segregation; Data from the Bolsa Família programme from 2005 to

2008 - provided by the Municipal Department for Social Promotion and Assistance. Map of the Urban Building Pattern - also provided by the city council; and finally, Dossiê de Jataí 2004 - in which we find an overview of the municipality, provided by the Superintendence of Industry, Commerce and Tourism.

The epistemological and specific reflections on urban geography were based on the works of Abreu (1994); Benko (1996); Carlos (1994); Castells (1983); Gottdiener (1997); Cavalcanti (2001); Deus (2002 and 2002a); Soares (2007); Gomes (1996); Lefebvre (1991 and 1999); Lojkine (1997); Marx & Engels (1989); Monteiro (1995); Santos (1993 and 1999); Santos and Silveira (2001); Seabra (2001); Silva (2003); Souza (2003 and 2003a) and Souza (1994). An understanding of the Centre-West region and its transformation - which involves the process of the arrival of the frontier in this area - was provided by the works of Estevam (1998); Graziano Neto (1985); Machado (1996); Martins (1975 and 1997); Steinberger (2003) and Teixeira Neto (2002). Also contributing to the process of regional understanding, specifically in the south-west of Goiás, were the following works: Agência Ambiental de Goiás (2003); Palacín & Moraes (1989); Palacín (1972); Palacín, Garcia & Amado (1995) and Teixeira Neto (1995). Although they exist in small quantities, the specific works on Jataí were fundamental to understanding the object of study: Campos Filho (2004); Melo (2003); Scopel (2002); Silva (2002) and the work that gave rise to this, the master's dissertation (SILVA, 2005). DATASUS (2004), GOVERNMENT OF GOIÁS (2004), IBGE (2000, 2002, 2003, 2004 and 2007) and UNDP/IPEA/Fundação João Pinheiro (2003) completed the statistical data necessary for the debates and reflections developed.

Based on the basic material, the work was structured into four chapters: Chapter 1 - Understanding the city of Jataí and its context; Chapter 2 - Building a necessary basis for analysing Jataí's urban space; Chapter 3 - Identifying urban segregation in Jataí and Chapter 4 - Access to urban land: the consolidation of segregated areas.

The first chapter presents the object of study, discussing the proposed research, the delimitation of the work carried out, its importance and justification. Jataí is characterised and contextualised within the subject matter, going back over its occupation process and arriving at the contemporary period. A temporal division is presented, based on key periods for the city, making it easier to understand the whole through its parts. At this point, the theoretical-methodological framework is also presented as a way of guiding the research.

The second chapter is made up of a discussion on the construction of the necessary bases for the development of the research. It characterises the urban fabric and discusses Jataí's place in the Brazilian urban network. The urban dynamics found in Jataí are discussed, taking into account the various agents present in the formation of this space. As a way of providing an understanding of the construction process of the cartographic base, the process of building the maps is presented, which constitute the substrate for discussing the data. At the end, reflections are made on the characterisation of the urban fabric presented.

The discussion of urban segregation in Jataí is at the centre of the third chapter. The "official division" made by the zoning law is presented. The segregated spaces are identified, as well as the issue of government assistance. The discussion closes with a proposal to look at the city differently, from the edge to the centre through a "peripheral gaze".

Finally, the last chapter proposes another element to answer the initial question. This section discusses access to urban land, correlating the price of land and the presence of urban voids with the elements in the previous chapters.

The answers reflect our view of the challenge we have set ourselves. They are not conclusive, but emerge in the expectation that this work will contribute to the process of building a fairer society, taking into account all the individuals involved in this process.

CHAPTER 1

UNDERSTANDING THE CITY OF JATAÍ AND ITS CONTEXT

"Your words were found, and I ate them; and your words were to me the joy and rejoicing of my heart..." Jeremiah 15:16

1.1 - DIRECTING RESEARCH: BRIEF CONSIDERATIONS ON URBAN STUDIES

Urban geography was initially a purely descriptive study. "The description of landscapes, buildings, works of art, the population and the activities carried out in the city formed the basis of the nascent urban studies" (ABREU, 1994, p.130). According to the same author, studies gradually evolved from simple descriptions to attempts to interpret the cityscape. However, the context and interrelationships between cities were still not taken into account, an aspect that cannot be left aside today when aiming for a real analysis, which aims for real, positive transformations in society. Subsequently, studies have analysed the relationships between urban spaces, going beyond the internal limits of the city. Urban geography developed and turned its attention to urban planning, as well as the use of urban land, once again focussing on the internal use of cities.

In the 1950s, Abreu (1994, p.132) comments, there was a theoretical-methodological break in urban geography. This endeavours to detach itself from neopositivism in the search for different methods of interpretation for its studies. The first moment in the production of urban geography, characterised by the description of reality, marked by classical positivism, gave way to the methods of neopositivism, paving the way for new approaches in studies. In this third stage, the multiplicity of theoretical and methodological references is one of the characteristics of Geography, which "attempts to explain the complexity and breadth of the urban phenomenon" Abreu (1994, p.133).

The modern city opens its doors for geographers to present themselves and carry out their studies using a variety of theoretical-methodological lines. The complexity of the city is currently before our eyes, so we must seek to answer its questions by making use, according to Abreu (1994, p.133), of the "diverse and multiple visions that the researcher has of the city". In this way, more consistent answers will be obtained by analysing the city from different perspectives and looking to the processes of transformation in society, which have the city as their place of realisation, to understand the urban scenario.

However, the use of the various "visions" available for analysing space should not be done in an empty, unfounded eclecticism. Santos (1999, p.51) reminds us that "space today is a system of increasingly artificial objects, populated by systems of actions equally imbued with artificiality, and increasingly tending towards ends foreign to the place and its inhabitants (our emphasis)". The urban landscape is so complex that we have lost the identity of places. The history of urbanisation shows a metamorphosis of the historic city in favour of modernity, with old forms taking on new functions to serve contemporary society. As a result, the place loses its essence and becomes alien to the community, which no longer feels part of this space. Here, in this reality, the geographer is inserted who, in the midst of so many variables, has the difficult mission of contributing to the understanding of space.

Seabra (2001, p.81) says that capital, after being generated,

> gains autonomy, and in its quest to enjoy greater profitability, decrees obsolescence on the one hand and produces the 'new' in the process of reproducing urban space on the other.

Urbanisation, which fragments space and builds places previously unimagined, means that the experience of this space is reduced to the private sphere. Based on this problem, the position adopted regarding the role of the geographer is the same as Seabra's (2001): the geographer must discern the representation of life, understand how life is represented, understand the representations of space and the interconnections with individuals within a materialist and dialectical vision.

Could the inhabitants of Jataí in the 1960s have imagined a space with the dimensions of today? How do today's inhabitants, living in the information age, see this space?

The expansion of Jatai's urban space does not simply express the spatial growth of society. Behind this visible aspect are unknown actors who leave their mark to be noticed, or merely disregarded by society: the old colonial mansion that becomes a museum, or another that has simply been demolished; the cobbled streets that express the "speed" of distant times, ignored by the necessary fluidity of modern society. How many other examples? Many have not been officially recorded.

The role of Geography, within this dialectical analysis, should be to perceive the encounters and mismatches[1] . To see beyond what is in the foreground. To unveil space. What is conceived and what is realised. The representations of this space, materialised in products, leading to a thought materiality.

[1] For a broader view of this analysis in Jataí, see SILVA (2005)

The vision of spatiality, made possible by geography, indicates the paths followed in the research. A vision of the spatial dimension of social reality, of the role of space in the production and reproduction of human life (CARLOS, 1994). The analysis of the urban space of the city of Jataí is carried out with a view to contemplating the various agents that make it up, seeking, above all, to reveal to "man" the space produced by "man", which is a complicated task, since it presupposes a look inwards.

In view of the above, we propose that the study of this urban space should not follow a rigid, ready-made path, but rather an analysis that starts from the urban landscape[2] , since it shows a vast horizon to be explored, arriving at the place[3] and analysing how it was produced. A path that leads to political and social analyses, contextualised in a so-called "post-modern" society, questioning and reflecting on the reality presented in the "geographical gaze", considering the action of capital and its relations in the construction of places. The dialectical perspective is presented as a guiding path for the work. The urban landscape, the starting point, is the result of political and social processes that have given it its current form.

Urban space is the realm of contradictions, and it is in this space that the various places are represented and transmit their message that constructs this space. Thus, some ideas of radical geography are pertinent, referring to its foundation, which lies in knowing the essence of facts - and not their appearances - linking

---

[2] Here we understand urban landscape according to Cavalcanti (2001, p. 14): "Urban landscape is the visible aspect of space, it is its formal, apparent expression. As a formal dimension, it expresses the content, the social relations that shape it. It is therefore historical, social and concrete [...]"

[3] Place here is understood "as an expression, materially spatial [...] as an element of a concrete totality, at the level of capitalist economic and social formation. [...] as a product of general determinations [reproducing] from specific historical determinations that differentiate places" (CARLOS, 1994, p. 39-40).

with historical and dialectical materialism in terms of the use of its characteristics and the political discourse engaged in science. In this way, the analysis of urban areas must take into account, above all, the political issues embedded in this context. The perspective of producing rational and objective knowledge by observing the real/historical as assimilated by critical geography is perfectly applicable to this analysis. It is "historical reason, materially determined, as opposed to the conception of idealism that defines reality as a product of absolute reason" (GOMES, 1996, p. 282). Here, reality is presented as a fundamentally social product.

Santos (1999) talks about two possibilities for working with geography. The first would be to consider fixes and flows, and the second territorial configuration and social relations. He considers that territorial configuration is the result of human production over time, which tends to negate natural nature as it replaces it with an "entirely humanised nature". Dialectics is present in the formation of space. There is an inseparability of a whole that is both solidary and contradictory. They are systems of objects and actions that must always be analysed together since they are part of the unique framework in which history takes place (SANTOS, 1999, p.51). Urban space is also denaturalised, as the place loses its original meaning to serve new interests .

It is considered that it is in praxis that man must demonstrate "the truth" (MARX & ENGELS, 1989). This statement, here taking the understanding of praxis as the transformative and conscious action of man, reveals the importance of tangible objects in order to analyse geographical space and thus not understand the production of real life as "[...] something separate from ordinary life, as something extra- and supra-terrestrial" (MARX & ENGELS, 1989, p.57).

In cities, "the truth", taking the previous statement as a parameter, can also be seen in the urban landscape. This landscape, crystallised in place, is the starting point for various analyses, including the formation of urban space. This space is a technical product - the product of productive forces - materialised in a place and inserted into a pre-existing fabric that will give it specific characteristics (SANTOS, 1999).

According to Seabra (2001, p.75) there has been a change in the quality of the urban phenomenon. At a certain point, the city as a form no longer supported its contents, giving way to metropolisation. Thus "[...] the metropolis is the contradictory synthesis of cities, and as such it holds fragments of what it involves: the city and the neighbourhoods". The process of spatialising social reproduction is metropolisation. It is the adaptation of the city to the rhythm imposed by reality. Development implies urban transformations: bridges, viaducts, transport, etc. and, according to the same author, it also leads to the emergence of neighbourhoods, which implies creating the conditions for structuring labour markets, product markets and, above all, the property market.

The historical foundations are torn apart, "...the city is a complex phenomenon... thinking about the city requires overcoming any linearity of reasoning" (SEABRA, 2001, p.78).

Historically, the city has produced a feeling of belonging to a community, but not everyone has this possibility, generating a struggle in the urban sphere in search of the right to "be a citizen or be nothing". A struggle in search of the co-right to be a citizen.

Analysing the course of time, Seabra (2001) highlights two movements in the urban space that would lead to a synthesis:

a) the first

> was that of property as an equivalent of value which, through the mechanisms of land rent, led to the formation of capital and the building and rebuilding of the historic city in terms of the city's products and works. The spatial segregation implicit in this process accommodated the contingents of urban workers in the city's neighbourhoods and made a clear distinction between the centre, the city and the surrounding area" (SEABRA, 2001, p.80).

b) the second

> It is capital itself, which thus gains autonomy, and in the search for where to enjoy greater profitability, decrees obsolescence on the one hand and produces the 'new' in the process of reproducing urban space on the other (SEABRA, 2001, p.81).

Based on these points, we realise that the remaking of the city has become big business, marked by the formation of a strong urban real estate sector that has the city as its field of investment and profit.

The history of urbanisation presents itself with a metamorphosis of the historic city in favour of modernity. Old forms take on new functions to serve society

contemporary. As a result, the place loses its essence and becomes alien to the community, which no longer feels part of this space.

Analysing the parts, the components of the analysed space, is a necessary task for understanding the whole. The parts are not restricted to the city under study, but also relate to its insertion in the state, Brazil and the world; in the social, economic and/or political spheres.

The world, as a result of human transformations, transforms man himself. The system of objects conditions actions, suggesting new actions that create new objects. This generates a dynamic transformation of space (SANTOS, 1999). In this dialectical relationship, needs are created that must be met for society to function properly, for its (re)production. The urban space of Jataí is one part of a wider whole, since it is part of a region that is the product of interregionalism (BENKO, 1996). Jataí's urban space is the result of human transformations and generates transformations in man himself. Space is thus a materialisation of man's experiences, perceptions and thinking (DEUS, 2002).

It reveals the society that produced it, because it comprises materiality plus the life that animates it, it is "an inseparable set of systems of objects and systems of action" (SANTOS, 1999, p.51). In order to understand the essence of this space, its forms must be unveiled. You have to extract the meanings of the object, going beyond a simple description of it (CASTELLS, 1983, p.78).

The urban landscape - the phenomenon, the appearance - contains elements of the social and historical context that produced it. The changes that took place in Jataí's urban space were the result of the needs of local society, a reflection of needs not necessarily created in the place in question. The objects implanted in Jataí with the modernisation of agriculture implied new social relations. The space previously identified with the local inhabitant tended towards "ends foreign to the place and its inhabitants" (SANTOS, 1999, p.51). The city has become the place of the rural dweller. They no longer harvest their food, but a wage that doesn't always allow them to have it. Marx & Engels, criticising German ideology in mid-1845, say that

> [...] the first presupposition of all human existence, and therefore of all history, is that men must be able to live in order to "make history". But in order to live, we must first of all eat, drink, live, dress and so on. The first historical act is therefore the production of the means to fulfil these needs [...] (MARX & ENGELS, 1989, p. 39). (Emphasis added).

Based on these assumptions, we realise that the production of these means marks the spatial configuration of Jataí. And, as a result, these means allow people to make history. History that is written by individuals who, in many cases, just want to be able to live a dignified life.

In turn, the history of the formation of Jataí is largely written by "moving the border". Space becomes a "place of disagreement" when the "other" is discovered (MARTINS, 1997). The "other", in this case, is the migrant, in contrast the individual without a financial situation that allows them to "make their history" in a less arduous way. This is where Jataí's urban landscape crystallises (Figures 1 and 2).

Figure 1: Jataí-GO: Urban area - partial view

Source: JATAI-GO (2004a): Jataí Dossier - Photo: Gênio Eurípedes

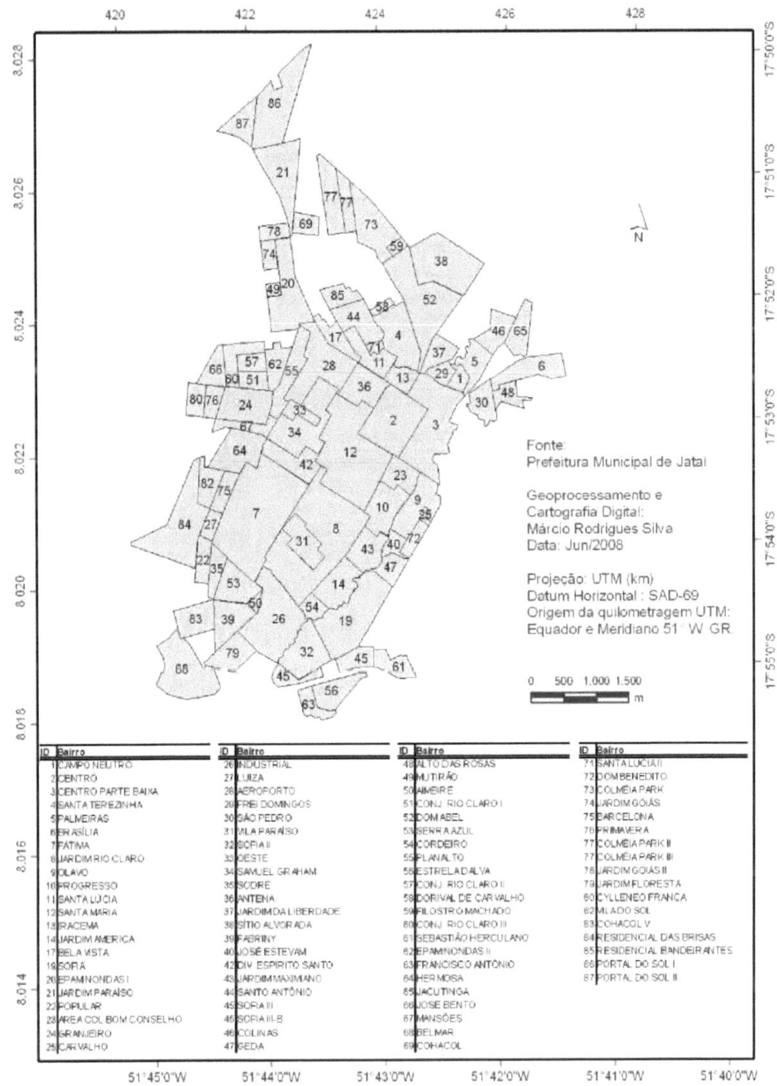

Figure 2: Jataí-GO: Urban area - 2008.

1.2 - THE TEMPORALITY OF THE RESEARCH: IMPORTANT PERIODS IN THE CONSTRUCTION OF JATAÍ

Santos (2001, p.23), when dealing with the problems of periodisation, warns that the important thing is to "[...] choose the key variables that, in each piece of time, will control the system of variables, this system of events that we call period" (emphasis in the original).

Although the task of periodisation is not so simple, it is fundamental. Separating periods with similar characteristics that drive the transformation process, in this case socio-spatial, is crucial to understanding the general context of the object of study and, from there, unveiling it.

Thus, "chunks of time" were delimited, defined by characteristics that interact, ensuring the movement of the whole, in order to define certain periods (SANTOS, 2001, p.24). Periods marked by instants of time occurring at a point in space, or simply by events (SANTOS, 1999, p. 115).

The choice of key variables is based on some classic time delimitations in terms of the Centre-West, such as that of Steinberger (2003, p.613), who highlights three periods, taking into account the process of productive transformation in this region:

> - The 1970s saw the beginning of the adaptation of the cerrado to soya cultivation, when the pioneers of this crop arrived in Mato Grosso do Sul and the first grain processing units were set up in Goiás;
>
> - The 1980s, emblematic of the expansion of soya and corn, dominated by commodities trading companies; and
>
> - The 1990s saw the consolidation of the grain-meat complex with the participation of international and national capital from large conglomerates in the south-east and south of the country. (Emphasis in original).

Estevam's (1998, p.168) observations point to similar periods.), there are similar periods: the first refers to the 1970s and 1980s, especially in the period 1975-1983, when "federal government resources earmarked for costing, commercialisation and investments in agriculture were voluminous; the second refers to the subsequent period, 1983-1990, when these resources suffered a gradual decrease; and the third is determined by an even greater fall in terms of rural credit, since "[....] at the beginning of the 1990s, rural credit suffered - at constant 1991 prices - a violent contraction, reaching levels close to those of the 1970s".

Thus, there are four significant periods for the municipality of Jataí-GO, and consequently for the formation[4] of its urban space:

- The first, which began with the occupation of the land in 1836, was marked by the first agricultural activities developed in the region. This lasted until the 1950s and 1970s, when extensive cattle ranching declined (MACHADO, 1996);
- The second refers to the arrival of the agricultural frontier and the beginning of the modernisation of agriculture in the 1970s;
- The third refers to the 1980s and 1990s, a period marked by the consolidation of the modernisation of agriculture, with the production of monocultures for export. A period of great dynamism in the urban space;
- The fourth period covers the current decade of the 2000s, when the traditional economic dynamics produced by the expansion of the agricultural frontier are exhausted, and the search for economic alternatives such as agro-industry (poultry, pigs, etc.), sugar and alcohol production and tourism stands out.

[4] The formation of urban space is understood as the formation of the visible aspect, the urban fabric itself, and the aspects that are not always perceived at first glance, namely the social relations of the city's residents, official and unofficial relations.

This division is not intended to establish rigid periods for this study, or even to define a fixed and finished temporal division that takes into account the different perspectives of geographical analysis. What we are trying to do is clarify some of the variables that are key to Jataí's socio-spatial configuration (Figures 1 and 2).

1.3- 0 RESEARCH CONTEXT

Nowadays, it's common to see an urban scenario marked by problems such as lack of employment, precarious transport, sanitation, health, etc. This situation is not unknown and is becoming increasingly noticeable as urbanisation grows. As Santos (1993, p.95) corroborates, "with differences in degree and intensity, all Brazilian cities display similar problems". The occupation of urban land currently requires serious control by regulatory bodies, such as local councils, and the role of public authorities in finding solutions to the urban scenario is fundamental.

We know, then, that instruments that can direct a growth process that caters for all social spheres are essential in the quest for city development. In one of these instruments, the Urban Master Plan implemented in Jataí, we see goals that, when achieved, will result in a fairer urban space and an improvement in the quality of life of the local population, since it provides for

> [...] ensuring a balance between demographic/economic growth and the quality of life and well-being of the population in the urban area [...] as well as [...] ensuring the general conditions for the development of production, commerce and services, and particularly for the full realisation of citizens' rights such as the right to health, education, basic sanitation, work and housing] to ensure the general conditions for the development of production, commerce and services and, in particular, for the full realisation of citizens' rights such as the right to health, education, basic sanitation, work and housing, public transport, safety, information, leisure and environmental quality and participation in planning. (JATAÍ-PLANO DIRETOR URBANO, 2001, p.1)

The situation in Jataí is no exception to Santos' (1993) observations: urban problems are present in its reality. This is where the geographer must act to help find solutions to the problems that often go hand in hand with the process of urban expansion.

As a way of informing public decisions on urban planning and also to support new studies aimed at contributing to the city's development, and not just its growth, this work analyses the internal structuring of Jataí, above all through population distribution by income level, in the predominantly post-1970 scenario.

To this end, bibliographical studies and fieldwork were used to gather data on the structuring of this space. Assembling a theoretical basis, necessary for decision-making, by cross-referencing information, which when interconnected reveals that what is visible is not always what is real, this work adds new elements to the discussion of Jataí's urban space, within the science of geography.

It is considered that work like this is of fundamental importance to provide data and information to the municipal government, as well as to society in general, with the aim of promoting debate, questioning and, above all, actions that promote "[...] authentic urban development, without inverted commas, [which] is not to be confused with a simple expansion of the urban fabric" (SOUZA, 2003, p.101).

1.4- 0 FORMATION PROCESS OF JATAÍ

The process of Brazilian urbanisation, which began in the 18th century with the first links between the so-called "fairs" (trade between farms), saw changes and adaptations in its form of action. The need to seek new horizons to fulfil national aspirations led to the gold rush which, in turn, opened up space for livestock farming. Similarly, the cultivation of tropical products (especially coffee production in the south-east) created subsidies for the establishment of industries. Thus, there are various aspects that lead to growing urbanisation, which would be a consequence of the whole process that society has experienced over time, leading to an occupation of the interior of our territory.

The first commercial stimulus that motivated bandeirantes and Jesuits to penetrate the extensive territory of Goiás was the capture of Indians for slave labour and catechisation. During these activities, gold from alluvial deposits was found and the news spread to Portugal. In 1722, the bandeirante Bartolomeu Bueno da Silva arrived in Goiás and located several gold-bearing streams in the region. From then on, Goiás was officially incorporated into Brazilian territory (PALACÍN, 1972).

The gold mining period in Goiás was brief and by 1822 production was almost non-existent (PALACÍN et al., 1995). The decline of gold made economic life very difficult for the mining population. The miners who remained in Goiás began to dedicate themselves to extensive livestock farming and subsistence agriculture. This new economic activity led to the expansion of the occupation of Goiás and an increase in population due to migratory flows from other states. At this time, the south-west of Goiás saw the emergence of new urban centres such as the city of Jataí (PALACÍN & MORAES, 1989).

After the historical changes seen in the national urbanisation process, today we are witnessing a space characterised by the actions resulting from a dynamic directly linked to the capitalist principles of profit and competitiveness. This space is therefore a product of the prevailing social structure, it is the concrete expression of the society that produces it.

The shape, and even the structure, of urban centres is always determined according to what is favourable to this or that group, with honourable exceptions. Urban space is not organised randomly, it is structured and the social processes linked to it express, by specifying them, the conditioning factors of each type and period of social organisation (CASTELLS, 1983, p.181-182).

The first families to settle in the south-west of Goiás saw attractions such as the "[...] abundance of good grazing land and easy access to property". Following the "path of the waters" and clearing the territory of Goiás, these settlements became true "[...] landowning estates that gave rise to cities such as Rio Verde, Jataí, Caiapônia, Quirinópolis, Mineiros, among others" (TEIXEIRA NETO, 2002, p.30). The municipality of Jataí (Figure 3), located in south-west Goiás, approximately 320 kilometres from the capital, emancipated in 1895, has its origins dating back to 1836, the year it was discovered

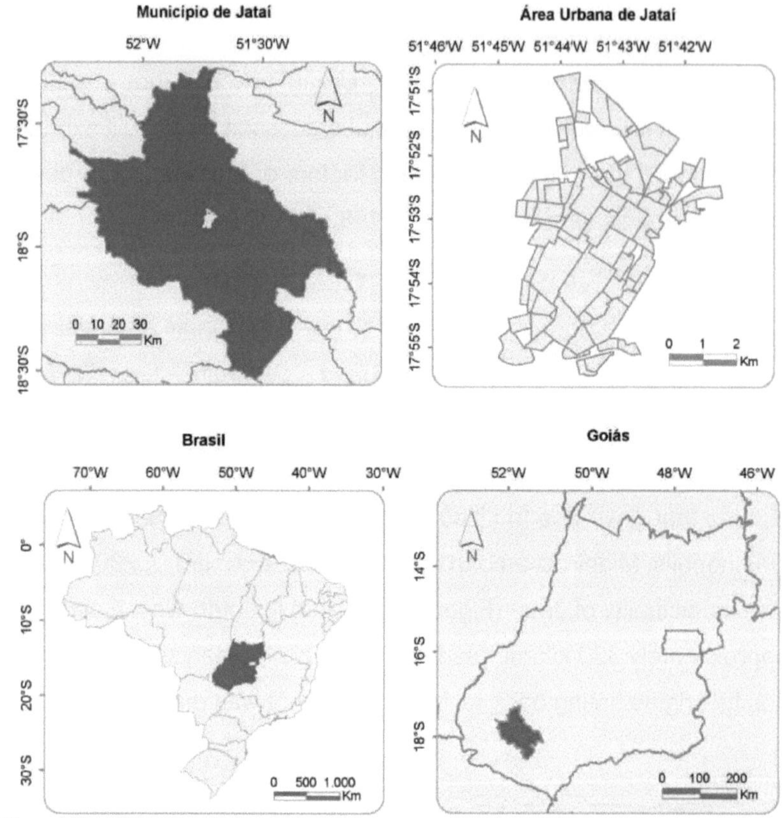

Figure 3: Location of the study area

of these lands. Its emergence is part of the movement to occupy the south-west of Goiás, since most of the towns in this micro-region were born out of traditional farming (TEIXEIRA NETO & GOMES, 1995). Like most of the Brazilian countryside, Jataí was born out of the courage and determination of men who, without any real reference, set off in search of new lands, whether for agriculture, livestock or other natural riches.

Its history is marked by the arrival of Francisco Joaquim Vilela and his son, José Manoel Vilela, in September 1836. After a long and tiring journey, they settled on the banks of the River Claro, where they planted crops and raised cattle. The first buildings were built on the banks of the Jataí stream and a later meeting with José Carvalho Bastos resulted in the division of the land, giving rise to the first possessions of the future municipality.

The beauty of the region impressed its new inhabitants and the hospitality of nature led to new migrations to the farm. The name Jataí comes from the large quantity of honey produced by the bee of the same name.

The District of Jataí was created in 1864, before the arrival of another pioneer: Serafim José de Barros. Another milestone would be, as in most Brazilian localities, the construction of the Mother Church of the Catholic Church in 1867.

According to Machado (1996), the main economic activity practised by migrants was extensive livestock farming, which lasted for more than a century, declining in the 1950s and 1970s.

From then on, a new period opened up that was quite significant for the urban structure of Jataí. The changes that have taken place raise questions that are part of a broad and complex reality. One side of this reality concerns the development of the Centre-West, influenced at the beginning of the 20th century by the expansion of coffee from São Paulo to Mato Grosso and the introduction of rail transport (Galindo and Santos, 1995).

More specifically, in relation to our area of study, we refer to the 1960s and 1970s when the foundations were laid for the introduction of modern fronts, which had a vigorous impact on the economy of the Midwest and its urban structure; a period in which the understanding of the productive structure and urbanisation of the Midwest lies (Guimarães & Leme, 1998, p.26).

In this sense, Galindo and Santos (1995) consider the Midwest to be a "frontier region" at national level, characterised by a strong expansion of agriculture and cattle raising. The national scenario, which led to major changes in the Centre-West, including those that took place in the south-west of Goiás, especially in the 1970s and 1980s, saw a period in which agriculture took on crucial importance in the expansion and diversification of exports. It was necessary to "produce food surpluses at reasonable costs and provide resources to finance the urban-industrial development of the dynamic centre of the national economy" (Mueller, 1992, apud Galindo and Santos, 1995, p.157).

The actions of the state were fundamental for the Centre-West to be occupied and to develop at significant rates from the 1970s onwards (Galindo and Santos, 1995). In this sense, Miziara (2000, p.284) comments on the state's participation in the arrival of the agricultural frontier in the Centre-West: "It was necessary for the state to provide other attractions, mainly in terms of credits and access to low-cost land, in order to change the calculation of private investment." This explains why the modernisation of agriculture in this region developed from the mid-1970s onwards, a period marked by "[...] specific policies for the region offering heavily subsidised credit". According to Miziara (2000), in the following decade, although there was a reduction in credit, the guarantee of minimum prices for the region was another form of subsidy.

After 1950 there were significant changes in the region's economic base. Farming became more commercial, with greater modernisation and diversification in the countryside; mining experienced new development fronts. These changes led to alterations in the secondary sector which grew from this base, taking advantage of the raw materials that originated there. Thus, "[...] the food products, timber and non-metallic minerals sectors are consolidated" (Galindo and Santos, 1995). In this set of transformations, Goiás gains a centrality that is not only geographical, defined by its coordinates - latitude and longitude, but is above all defined by the role that its territory plays in relation to the country as a whole (TEIXEIRA NETO, 2002, p.12).

1.4.1 -BUILDING THE BORDER

The emergence of Jataí was initially part of the advance of the "expansion front" and later of the "pioneer front"; more recently it was marked by the advance of the agricultural frontier in the 1970s.

We can understand the first moment, of the "expansion front", as "the movement in which economic life is not structured primarily on the basis of relations with the market, but supplies products for commercialisation, and cannot therefore be classified as a natural economy" (FERREIRA, 1988, p.41).

The "pioneer front" is set up as an economic enterprise. It would be "[...] a way of extending the frontier, incorporating new areas into the productive system, where the social relations that underpin it will be reproduced" (Ferreira, 1988, p.41), constituting the second moment of the occupation. This "[...] expresses a social movement whose immediate result is the incorporation of new regions into the market economy. It presents itself as an economic frontier" (Martins, 1975, p.45).

Here we refer to the border as

> [...] the process of occupying a space that is considered empty. The emptiness can be demographic, economic or legal, and the space can be found in the Amazon rainforest, the savannahs or anywhere in Brazil (FERREIRA, 1988, p.38).

The area in question is the city of Jataí, which is part of the post-1964 reality of the centre-east and, consequently, the reality of Goiás in the context of national integration policies. This reality gave

Goiás the role of "agricultural frontier, producer of food and raw materials [...]" (Barreira, 2002, p.178).

According to Teixeira Neto & Gomes (1995, p.7)

> the new model of occupation and settlement was no longer as spontaneous as it was at first, with the arrival of the first inhabitants, but was commanded and organised by capital and obeyed the goals and priorities of the national and international capitalist bourgeoisie. Land ownership was no longer public, but already duly appropriated, privatised and individualised.

According to Barreira (2002, p.177), "in the already occupied areas [of Goiás territory], there was a redefinition and deepening of the more articulated economic processes". These areas suffered "[...] an intensification of activities, in a more capitalist way". This space was seen by individual capitalists as an "agricultural frontier zone", offering "[...] the possibility of altering one of the variables - social, geographical or economic - according to their interests [...]" (Miziara, 2000, p.283). It is in this area that we can see the influence of the arrival of the frontier, especially the agricultural frontier, which alters the local economy:

> The south-west of Goiás, as well as the municipality of Jataí, has had traditional livestock farming as its main economic activity since the beginning of its settlement. From the 1970s onwards, it gradually incorporated "modern" forms of agricultural and livestock production (MELO, 2003, p.78).

Barreira (2002) analyses the evolution of the occupation of Goiás through regional incorporation by the pioneer fronts. According to the author, this analysis aims to demonstrate the redefinition of the current territory with the development of the fronts. This redefinition was already highlighted by Borges (1990), who analysed the arrival of the railroad in Goiás; a strong element that drove regional development: "the numerous highways built in the early 1920s [...] meant that the radius of expansion of the modernising process went beyond the limits of the railroad region" (Borges, 1990, p.108).

Figure 4 reproduces the map of the expansion of pioneering fronts in the 20th century drawn up by Barreira (2002). It shows the incorporation of south-west Goiás into the national market in the 1920s. It also shows the direction of the occupation originating in Minas Gerais, which was later influenced by the state capital. Looking at this map gives an overview of Melo's (2003, p 42) description of the south-west of Goiás at the beginning of the 20th century.

> Compared to the south-east of Goiás, in the early years of the 20th century, the south-west experienced a "slow time". Socio-economic life moved at the same pace as the means of transport, in tune with the trotting of the horse and the "singing" of the ox cart. But even in its particular slowness, it was not resistant to the new, the modern. These were, however, the dreams of a significant part of the society that was forming in Rio Verde, Jataí, Mineiros, Rio Bonito (Caiapônia) and Santa Rita do Paranaíba (Itumbiara), which launched cattle from the southwest into the markets of Minas Gerais and São Paulo, through hard and painful journeys.

Figure 4: State of Goiás: Expansion of the Pioneer Fronts in the 20th Century

Estevam (1998, p.189), speaking about the productive structure and urbanisation in Goiás, stresses the importance of understanding the socio-economic transformations that have taken place in the state within the spatial framework of the Brazilian Midwest. The changes that have taken place in its socio-productive structure,

> [...] especially in the post-1970 context, the reduction in the necessary working time/production period changed regional socio-economic relations, especially in the centre-south of the state (ESTEVAM.1998, p.170).

This region "articulated itself, as the 'frontier of capital', to the hegemonic space of national capitalism, with São Paulo as its dynamic centre". In this sense, Borges (1990, p.12) talks about the expansion of the economic frontier from the south-east of Brazil, which "[...] established or redefined the regional division of labour, which would serve as the basis for the growth of the entire process of urbanisation and industrialisation that was implemented in the country [...]".

Barreira (2002, p.171) analyses the structural changes that took place in the south-west of the state after the mid-1940s:

> [...] the regional structure changed with the introduction of commercial agriculture (rice and cotton) in municipalities such as Jataí, Rio Verde, Santa Helena, Quirinópolis, Paraúna, Mineiros and Santa Rita do Araguaia. With the arrival of soya in the 1970s and the consequent modernisation of agriculture, unprecedented transformations were unleashed. Today, this region concentrates the largest agricultural production, with important agro-industries and highly commercial agriculture.

The set of spatial changes described is the basis for verifying the significant effects of the "expansion of the frontier" in Jataí. It is worth highlighting the words of Graziano Neto (1985) when he talks about the modernisation of the

This is the reality of the south-west of Goiás and of Jataí. In the municipality of Jataí, there is a marked presence of the modernisation of agriculture, but

> [...] modernisation cannot be confused with development, because development means a process of socio-economic transformation that brings about an improvement in the general well-being of the population, a rise in the standard of living as a whole" (GRAZIANO NETO, 1985, p.78).

Commenting on the effects of the modernisation of agriculture, Brumer (n.d., p.184) says that "the [population] transfer from the countryside to the city does not mean [...] greater well-being for the workers, who are practically 'abandoned'", which translates into growing unemployment and/or underemployment in urban areas, causing serious social problems. This is the presence of modernisation, but it doesn't necessarily imply development that goes hand in hand with it.

According to Lefebvre (1991, p.4), cities are "[...] centres of social and political life [...]" and so they crystallise the changes that have taken place in these spheres in the urban space. These changes, fundamentally due to the modernisation of agriculture, were crystallised in Jataí and make up the space analysed here.

1.4.2 - THE URBAN RESPONSE

The scenario of the major regional changes that have taken place in the Midwest is marked by a high rate of urbanisation in the period from 1970 to 1991, as Galindo and Santos (1995) verify by analysing data from the 1991 Census. In 1970, 48 per cent of the total population lived in urban areas. In 1991, this percentage rose to 81.3 per cent. These changes have resulted in

> [...] both the transformations in production relations and the technological process adopted by agriculture, fundamentally geared towards the production of export crops, whose mechanisation has reduced the absorption

> capacity of the contingent working in the countryside, forcing the rural population to seek employment in urban areas (Galindo and Santos, 1995, p.179).

In 1980, the state of Goiás had an urbanisation rate of around 62%, rising to 76% in the following decade in a clear urban growth (MARTINE, 1994).

Deus (2002) discusses the effects of the arrival of the agricultural frontier in Goiás. According to the author, this model of agricultural production, which uses specialised or unskilled but temporary labour on large properties (latifundios), caused a rural exodus and, consequently, an intense process of urbanisation. As a result, the southern mesoregion of Goiás, made up of the micro-regions of Sudoeste de Goiás, Vale dos Bois, Quirinópolis, Meia Ponte, Pires do Rio and Catalão, includes "[...] the largest non-metropolitan centres in Goiás, with the exception of Anápolis. Rio Verde, Itumbiara, Jataí and Catalão stand out, driven by mechanised agriculture, producing grains such as soya and corn

In this sense, Martine (1994) agrees with us when he observes that one of the characteristics of this type of expansion - based on capital-intensive agriculture, where soya plays a major role - is precisely that few people are employed in rural areas, although it can generate jobs in the formal and informal sectors of urban areas.

With regard to the urban space of Jataí, an understanding of the frontier is based on the statements made by Martins (1997):

> [...] the border is essentially the place of otherness. That's what makes it unique. At first glance, it's the place where those who for different reasons are different from each other meet [...]. But the conflict means that the border is essentially, at the same time, a place of discovery of the other and of disagreement. (MARTINS, 1997, p.150). (Emphasis in original).

Jataí becomes a stage for change. What was tangible is

transformed at a previously unknown pace. New needs are created. New characters appear in the space and demand their place in society and in the city. For some, their new place is the noble neighbourhood, with its imposing buildings. For others, the majority, it's a place that wasn't dreamt of, it's the periphery, it's low-paid labour.

The same space of change is the place of "discovering the other" and recognising the benefits it brings, which we can see from the economic advance: "The most prominent period in the founding of agricultural trade and service companies [in Jataí] occurred, above all, in the 1980s, the historical moment of the expansion of soya cultivation in the municipality [...]" (Melo, 2003, p.115).

> The occupation of the Midwest, marked fundamentally by the presence of agriculture and livestock, is reflected in the process of urbanisation of this space, both in terms of services and consumption, transforming several cities that were characteristically agricultural until the 1960s "[...] into urban centres with relatively autonomous dynamics of income and employment expansion" (Galindo and Santos, 1995, p.166).

Analysing Graph 1, we can see the population growth in the city of Jataí. In 1970 there were around 40,000 inhabitants; a figure

Graph 1 - Jataí-GO: Population evolution - 1910 to 2007

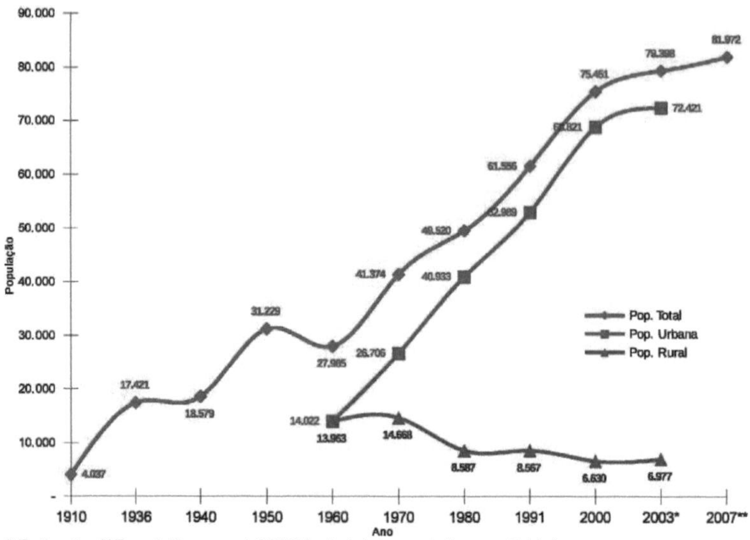

* Estimate. **Population count 2007 (only total population available)
Source: IBGE: Demographic census, 2000/Population estimates, 2003/Twentieth century statistics, 2004/Population count 2007; MELO, N. A. (2003).

which reached more than 75,000 in 2000 and almost 82,000 in 2007. This allows us to question the dynamic urban framework of the period in question. According to Melo (2003), the growth of the urban population between 1970 and 2000 continued to increase, since in the 1970s people living in the city were already in the majority.

There was a growth of approximately 157%, to the detriment of the rural population, which decreased in the three decades considered, especially between 1970 and 1980 (a 55% decrease). This places Jataí among the two cities in south-west Goiás with an urban population of over 90% (Table 1).

Table 1 - Goiás: Municipalities with predominantly urban populations - 2000

Municipalities with an Urban Population > 90% of the total
Aguas Lindas de Goiás
Anápolis
Aparecida de Goiânia
New Falls
Goianésia
Goiânia
Inhumas
Itumbiara
Jataí
Luziânia
Santa Helena de Goiás
New Gama
Planaltina

Santo Antônio do Descoberto
Senador Canedo
Trinidad

Source: Adapted from AGÊNCIA AMBIENTAL DE GOIÁS, 2003 (GeoGoiás 2002)

Since the beginning of Jataí's occupation, its population dynamics have been significant, for the reasons already explained. From 1940 onwards, the rate of growth remained similar to the initial rates. However, in the period from 1950 onwards, the motivation is no longer the same as it was in the first occupations, since "[...] the large rural migratory flows have been cancelled out".

directed towards the expanding agricultural frontier, consolidating existing centres in the southern part of the state" (AGÊNCIA AMBIENTAL DE GOIÁS, 2003, p.1).

The 1960-1970 and 1991-2000 intervals stand out in this table.

The first was the modernisation of agriculture as the driving force behind growth, and these were the years with the highest percentage growth in the most recent period (48%). The second was the consolidation of monocultures for export, marked by new dynamics in the urban space, including the large number of commercial, industrial and service establishments set up in Jataí. In the 1960s, the urban population began to outnumber the rural population, but still in a timid way. It was in the following decade that this trend was consolidated.

Population growth continues at a steady pace, allowing for similar estimates for the current decade. It is worth emphasising that the municipality's population did not fall in any period. The difference between 1950 and 1960 is due to the emancipation of the Serranópolis district in 1958.

Today, the city has changed mainly as a result of the

process of capital reproduction imposed by the arrival of the agricultural frontier.

> Agricultural production developed at the same time as the mechanisation of the production process, the advance of salaried work, the incorporation of specialised work, the rural exodus, the "technification" of space, the growth of the city, as well as a series of socio-cultural and economic changes (MELO, 2003, p. 85).

According to Martine (1994, p.5) the 1980s can be considered one of the most remarkable of the 20th century: "The most notable transformations, of course, took place in the political and economic sphere. But the population area was also characterised by extremely important changes". This same decade plays a unique role in our analysis. It was during this decade that the arrival of the agricultural frontier in the municipality was consolidated (it was also the period with the highest volume of resources from rural credit for the Centre-West region - Table 2) bringing

Table 2 - Brazil: Regions' participation in rural credit from 1966 to 1999/00 (selected years).

	Brazilian Regions (%)				
	South East	South	Centre-West	North/Northeast	
1966	47	30	-	23	100
1970	45,6	31,8	6,5	16,1	100
1975	35,7	38,2	10,1	15	100
1980	34,1	35,8	10,5	19,6	100
1985	26,2	41,6	16,3	15,9	100
1985/86	28,2	37,6	18,1	16,1	100
1986/87	26	39,1	18,4	16,4	100
1987/88	24,3	42,5	22,4	10,3	100
1988/89	21,1	34,5	32,9	11,3	100
1989/90	31,5	41,1	17,1	9,6	100
1990/91	26,7	43,3	19,3	10,7	100
1991/92	21,5	47,3	21,5	9,6	100
1992/93	21,7	48,9	21,6	8,4	100
1993/94	18,3	41,5	28	11,1	100
1994/95	26,6	38,9	21,1	13,3	100
1995/96	24,3	38,5	19,2	17,9	100
1996/97	30,7	37,7	20,1	11,4	100
1997/98	30,6	37,1	18,5	12,9	100
1998/99	29,6	39,8	17,7	11,9	100
1999/00[1]	28,7	42,1	18,4	10,8	100

Source: MELO, 2003, p. 69.[1]Preliminary data

with it a new dynamic for the urban sphere as a result of new demands from the rural sphere, as well as an increase in population, specifically marked by a decrease in the rural population compared to the previous decade.

1.4.3 - CHANGES IN THE FOUNDATIONS OF URBAN SPACE

From the 1970s onwards, changes began to take place in the structure of land use in Jataí and other neighbouring municipalities, with the introduction of modern agriculture for the production of mechanised annual crops and exports, especially soya, which had repercussions not only on the rural scene, but also significantly on urban areas.

This reorganisation of rural and urban space is part of the process of reproduction and expansion of capitalism within the Brazilian territory (MACHADO, 1996), imprinting the mark of an economic system which

> [...] it tends to expand territorially and sectorally, driven by technical innovations that increase the productivity of labour, allowing the growing generation of a surplus of production that frees workers from service in the countryside at the same time as it manages to supply an increasingly urbanised population (SANTOS 1998, p.10).

As a result, the population growth rates of the south-western region of Goiás increased, even surpassing the state's growth rates by more than 10 per cent in the 1980s. A large number of farmers, mainly from the south, motivated by the diversity of credit lines available, as well as subsidised interest rates (ESTEVAM, 1998), arrived in the region bringing intensive soil mechanisation and the expansion of monocultures, the most significant being soya and maize. During this period, and as a result of the migration process linked to the occupation process, the urban population surpassed the rural population (MACHADO, 1996).

The result of these events could not have been different:

urban areas began to receive contingents of workers "freed up" by the modernisation of agriculture. This has had a significant impact on the cities, especially Jataí, which has seen a significant rate of urbanisation.

Currently, Jataí stands out as a grain producer and is starting out in the sugar-alcohol sector. Its population is 86,447 (IBGE - Estimativa da População, 2009), representing 1.51% of the population of the state of Goiás in 2000, with an urbanisation rate of 91.21% (PNUD/IPEA/FUN DAÇÃO JOÃO PINHEIRO, 2003). Its annual growth rate in the period 1991-2000 was an average of 2.17%, and the urbanisation rate grew by 6.86% in the same period.

With a longevity index of 71.5 years, the city had a human development index (Table 3) of 0.793 in 2000, with education contributing 42.9% to the 10.75% increase (from 1991 to 2000) in the municipality's HDI. Although the indices show a slight improvement over the period in question, the figures are still worrying: the poverty index in 1991 was 31.16%, rising to 20.25% in 2000; according to these figures, 9.41% of the population lives on government transfers (pensions, retirement pensions, assistance programmes, etc.), compared to 6.57% in 1991. In this period, inequality remained practically the same, with a drop in the Gini index from 0.60 in 1991 to 0.59 in 2000 (UNDP/IPEA/FUNDAÇÃO JOÃO PINHEIRO, 2003).

Table 3 - Jataí-GO: Human Development

	1991	*2000*
Municipal Human Development Index	0,716	0,793
Education	0,775	0,874
Longevity	0,687	0,776
Income	0,685	0,728

Source: (UNDP/IPEA/FUNDAÇÃO JOÃO PINHEIRO, 2003)

We can see major changes in the municipality of Jataí if we compare it with its period of origin. Over the years, the municipality's agriculture (Table 4), livestock (Table 5), industry and commerce (Table 6) have developed greatly, making it a standout in the state. With a dynamic economy, Jataí accounts for 55.05% of the state's agricultural production.

Table 4 - Jataí-GO: Agricultural Production and Planted Area - 2000-2001

Products	2000		2001	
	Ton.	ha	Ton.	ha
Rice	9.450	4.500	10.800	4.500
Beans	2.263	1.886	3.382	2.355
Sunflower	2.329	1.941	4.020	2.680
Maize	357.486	100.529	400.585	90.021
Soya	462.462	157.300	483.542	164.470
Sorghum	35.847	23.898	53.801	20.693

Source: SEPIN/SEPLAN apud JATAI-GO (2004a).

safrinha maize in the state of Goiás and more than one fifth of total sorghum production (JATAÍ-GO, 2004a).

Table 5 - Jataí-GO: Livestock Production 1998-2000

Products	1998	1999	2000
Poultry (heads)	1.865.100	2.366.600	2.372.000
Eggs (1,000 dozen)	352	359	375
Cattle (head)	392.338	411.752	377.343
Milk (1,000 litres)	41.174	43.198	45.448
Milked cows (head)	30.900	32.419	34.120
Pigs (head)	23.350	26.400	29.920

Source: SEPIN/SEPLAN apud JATAI-GO (2004a).

Table 6 - Jataí-GO: Economic Activities Registered with the Secretariat of Fiscal Management in the period "Up to 1959" to 2000 up to 1959" to 2000

Decade	Trade*	Industry*	Provision of Services*	Total*	Difference
Until 1959	1	0	3	4	0
1960	4	2	7	13	9
1970	19	16	54	89	76
1980	189	52	200	441	352
1990	842	70	967	1879	1438
2000**	958	120	1430	2508	629
Grand Total	2013	260	2661	4934	

'accumulated figures ** up to 2004

Source: Fiscal Management Secretariat. Report on registered economic activities 2004.

There are industries such as Coinbra - a multinational that industrialises soya; Nestlé, which pre-condenses milk; Perdigão, which industrialises poultry and its derivatives; as well as clothing industries and others. These industries are supported by the

Agroindustrial District II[5] , which is located in the centre of the city. at the junction of BR 364 and BR 060 and which, according to information from the local administration (JATAÍ-GO, 2004a), is ready to receive industries from all over the country and has proved to be a strong attraction for the business sector.

Occupying a strategic geographical position, Jataí (Figure 3) is a connection point between the South-North and East-West regions, and is crossed by the BR-364 motorway. It is a strong point for industrial, commercial and service development, as noted by Teixeira Neto & Gomes (1995, p.5):

> The south-west of Goiás, with the city of Jataí as its industrial, financial and cultural hub, due to its privileged location and geographical position, equidistantly centralises a large part of the production of the vast area of the Centre-West of Brazil. [...] the potential fertility of its land, combined with the region's centuries-old tradition of farming and cattle raising, and the major motorways, are other valuable factors in regional integration.

The town has four higher education centres, two of which are federal: UFG (Campus of the Federal University of Goiás), with eighteen courses, and IFET (Federal Institute of Education, Science and Technology); one state: UEG (State University of Goiás); and one private: CESUT (Jataí Higher Education Centre). It has educational units such as: EMPPA (Escola Municipal Profissionalizante - PROJETO ABELHA), with more than 14 courses; SENAC (Serviço Nacional de Aprendizagem Comercial) and SEBRAE.

In the health sector, the municipality has three hospitals with a total of 253 beds in 2004 (JATAÍ-GO, 2004a), 238 in 2005 (SEPLAN, 2005) and 201 in 2008, 176 of which are SUS beds (DATASUS 2008).

The municipality's infant mortality rate (Table 7) fell by 30.53 per

[5]ZI-II - see item 3.1.3- URBAN ZONES: THE OFFICIAL DIVISION

cent in the period 1991-2000, reaching 17.77 (per thousand live births) in 2000, compared to 25.58 in 1991.

Table 7 - Jataí-GO: Longevity, mortality and fertility indicators, 1991 and 2000

	1991	2000
Mortality up to 1 year of age (per 1000 live births)	25,6	17,8
Life expectancy at birth (years)	66,2	71,5
Fertility rate (children per woman)	2,7	2,6

Source: (UNDP/IPEA/FUNDAÇÃO JOÃO PINHEIRO, 2003)

As a way of diversifying the municipality's economic activity, Jataí also invests in tourism (Figure 5 and Appendix A). The media publicise it:

> The city council built three ecological parks which were of fundamental importance in helping Jataí to move away from being a basically agricultural city, which it had always been, to take on a more prominent role (CAMPOS FILHO, 2004, p.2).

Figure 5: Jataí-GO: Vale do Paraíso tourist centre - Bonsucesso Lake Source: JATAI-GO (2009) - Photo: Amarildo Gonçalves

A reference to the words of Lefebvre (1999, p.33) is in order here. In Jataí, the "metamorphosis produced by the capitalist machine" is clear, in which the city is at the same time "the place, the instrument, the dramatic theatre" of this transformation.

In this sense, actions in the urban space are seeking new paths. "After the construction of the ecological parks, a new joint effort is now beginning, in search of alternatives to generate jobs and income for the population" (CAMPOS FILHO, 2004, p.4).

The fight for development, as part of a kind of "war of places" (SANTOS, 2001), advertises the tranquillity of the countryside (although not so much) in order to attract investment and foreign exchange: "Without the violence and problems of the big cities [...] Jataí is a feast of attractions for thousands of people looking for a paradise away from the worries of everyday life" (CAMPOS FILHO, 2004, p.7).

Going back to what Lefebvre (1999, p.33) said, it is to be hoped that the "theatre" is not too dramatic, and that this is a way of bringing development to the south-west of Goiás. What is worth highlighting is the dynamism present in a city that, for a long time, was based on traditional farming and cattle-raising, but which today, inserted into a globalised world, tends towards ends that are foreign to the place and its inhabitants (SANTOS, 1999, p.51).

This study is one of the elements needed to promote, assist, correct and/or reformulate local planning, thus directly influencing the improvement of society's quality of life, providing a new urbanisation, as well as a "re-urbanisation" when it becomes necessary, seeking to meet the most basic needs of the population.

It is through a partnership between researchers and public authorities that the transformations that have been so longed for and hoped for, especially in relation to "our home" - *the city* - will be driven out of a theoretical realm and materialise in a practice that meets the interests of the majority of the population (which is so forgotten today).

Unilateral actions don't have the same strength as a joint struggle, which involves interconnected actions from the different sectors of society. It is crucial to harness the potential that exists in the various spheres of society so that, through firm and coherent debate, urban development flows in the best possible way. We need to think, reflect and try to understand the local urban phenomenon, with a view to finding solutions that we don't have (CARLOS, 1994).

CHAPTER 2

BUILDING A NECESSARY BASIS FOR ANALYSING JATAI'S URBAN SPACE

"Unless *the Lord builds the house, those who build it labour in vain; unless the Lord guards the city, the watchman watches in vain." Psalm 127:1*

It is necessary, within a more complete urban analysis, to consider, on the one hand, the urban as a general condition for the realisation of the process of capital reproduction. On the other hand, to consider it as a product of the process, the result of contradictions, of conflicts between what is demanded by capital, its needs, and the needs of society (CARLOS, 1994). This points to a city built with contradictory elements. Wealth and poverty. Joy and sadness. Closeness and distance. Elements originating from the man who builds it, who projects onto a place the society of which he is a part (LEFEBVRE, 1991).

Analysing these elements and the relationships resulting from their interactions is a complex task due to the complexity of the context being worked on. In agreement with Carlos (1994, p.13), "the path to knowledge is fascinating, but arduous and painful, a process that cannot be predicted in its entirety". Various processes influence the city: economic, social, political and cultural. By shaping the urban space, they have shaped the city. Man and human groups have built, and are building, space, using their own rhythms (LEFEBVRE, 1991).

The fascination was sometimes realised with the results already achieved, when these opened up new questions, new paths, requiring new research, sometimes arduous and painful.

Part of the difficulty arose from the temporal diversity of the data used in the research. Data from various dates was used, requiring approximations and projections in order to analyse reality more faithfully. This difficulty was caused by the unwillingness of official bodies to make the data available. Obstacles were encountered, especially at local level, either due to the lack of records at these organisations, the lack of time tracking, changes in storage methods and even the loss of information (which occurred at the town hall due to a change in administrative software). However, no effort was spared to minimise this shortcoming, given that the results closely reflect the reality found in Jataí.

Understanding some of the elements that make up Jataí's dynamics is the necessary basis for a more solid analysis. The following elements include factors that directly or indirectly influence the urban structure in question.

2.1 - JATAÍ'S PLACE IN THE BRAZILIAN URBAN NETWORK

In order to understand the city of Jataí, it is pertinent to briefly discuss its network, considering the current dynamics that are rapidly transforming the urban and regional reality, and understanding that

> the centrality of the urban fact, which refers both to the organisation of the immediate territory and to the articulation of economic space in extended territories, permanently generates and expands new regions and networks of localities in its areas of influence, also redefining new urban centres that command ever wider areas of production and consumption (MONTE-MÓR, 2004, p.3).

From the perspective of the proposed analysis, Jataí's insertion into the urban network of Goiás takes place according to its main activities. Today, Jataí is an agriculturally based city, with an emphasis on grain production.

> Agribusiness in the region is one of the most expressive at a national level, due to the use of cutting-edge technology, especially in the meat and grain complex, with Jataí being one of its main exponents. This modernisation, especially in agriculture, has led the municipality to consolidate its position as the largest grain producer in the state of Goiás and to become one of the largest grain producers in the country" [...]. Its potential is growing every year, with the highest grain yield per hectare/year planted, the highest corn production and the ninth largest soya producer in the country. Beef and dairy cattle breeding is at its best in terms of genetic improvement, and it owes nothing to other states in this area. (SEPI_AN, 2005, p.61).

According to SEPLAN data (Table 8), Jataí ranks second in agriculture in the state. It ranks ninth in industry, eighth in services and seventh in GDP.

Table 8 - Ranking of municipalities by sectors of activity - 2000

Agriculture	Industry	Services	GDP
Rio Verde	**Goiânia**	**Goiânia**	**Goiânia**
Jataí	**Anápolis**	**Anápolis**	**Anápolis**
Crystal	**Catalan**	**Aparecida de Goiânia**	**Rio Verde**
Minas Gerais	**Itumbiara**	**Rio Verde**	**Aparecida de Goiânia**
Chapadão do Céu	**Aparecida de Goiânia**	**Senador Canedo**	**Catalan**
Luziânia	**Rio Verde**	**Itumbiara**	**Itumbiara**
Montividiu	**Minaçu**	**Luziânia**	**Jataí**
Caiaponia	**Luziânia**	**Jataí**	**Luziânia**
Quirinópolis	**Jataí**	**Catalan**	**Senador Canedo**
Santa Helena de Goiás	**Trinidad**	**Valparaíso de Goiás**	**Minaçu**

Source: SEPLAN, 2004

As a dynamic region in south-west Goiás, Jataí polarises the municipalities of Caiapônia, Perolândia and Serranópolis, and is classified as a "Zone A Centre[6] ", Table 9 (IBGE, 2007b). Jataí, Rio Verde and Mineiros are the main cities in this micro-region, in a trend already observed by Monte-Mór (2004), when he highlighted the

> proliferation of small and medium-sized urban centres articulated in new socio-spatial arrangements, seeking institutional innovations for the provision and management of municipal and/or micro-regional services. At this level, local urban systems emerge in which the various centres play complementary roles and develop actions in conjunction with each other. Thus, one or more centres can be privileged spaces for health and education services,

[6] This level is made up of smaller cities with operations restricted to their immediate area; they perform elementary management functions. Levels 5 and 6 of territorial management predominate (IBGE, 2007b).

while others bring together support services for the productive bases of the micro-region (MONTE- MÓR, 2004, p.12).

Table 9 - Selected urban centres: centrality levels according to REGIC-2007

	City centre				
	Miners	Jataí	Rio Verde	Goiânia	
Denomination of centrality	Zone A centre	Zone A centre	Centre Subregional A	Metropolis	Variation
Level of centrality	9"	9°	7°	3°	Del a11
Level of centrality of Federal Management	6"	6°	6°	4°	From 1 to 6
Level of centrality of Business Management	7"	5°	6°	3°	From 1 to 8
Level of diversity in trade activities	4°	3°	3°	1°	From 1 to 5
Level of diversity in service activities	5"	4°	4°	2°	From 1 to 5
Level of centrality in financial activities	6°	6°	5°	5°	From 1 to 8
Level of centrality in television networks	*	4°	4°	2°	From 1 to 5
Level of centrality as territorial management centres	5°	4°	5°	3°	From 1 to 6
Level of centrality in undergraduate teaching	6°	5°	5°	2°	From 1 to 6
Level of centrality in health services	6"	5°	5°	2°	From 1 to 8

*Unclassified

Source: IBGE, 2007b

With courses in Agronomy and Veterinary Science, as well as Geography and Biological Sciences, the municipality's agricultural base has a qualified workforce that graduates every year. Whether in activities directly or indirectly linked to the maintenance and development of the local economy, Jataí stands out as a training centre in south-west Goiás, generating flows of students from the surrounding area to the city in search of professional qualifications. The main flows come from Mineiros, Caiapônia, Serranópolis, Caçu, Cachoeira Alta, Aparecida do Rio Doce, Santa Rita do Araguaia, Iporá and Alto Araguaia-MT (SEPLAN, 2005).

Another important element in the urban network of Goiás is the roads. The roads present in municipality are crucial for the flows from this point, especially the motorways.

> Road access to the region is via several highways: BR-060, BR-364, BR-158 and GO-184. The BR-364, which is vital for the municipality, is responsible for transporting its grain production to the main Brazilian ports, as it connects Jataí to the Anápolis Dry Port and the Tietê-Paraná Waterway at the Port of São Simão. The city also has a modern airport with an asphalted runway measuring 1,494 metres in length and although there is no regular flight line, there is a considerable movement of private aircraft and air taxis (SEPLAN, 2005, p.62).

The economy, based on agriculture, livestock and agro-industry, has made it possible to have "the second highest added value in the agricultural sector in Goiás, behind only Rio Verde, and the sixth highest GDP among Goiás municipalities, corresponding to R$824.10 million" (SEPLAN, 2005, p. 62).

The garment industry in the city is another element that boosts Jatai's economy, absorbing a significant amount of labour. According to SEPLAN (2005), one of the main companies in this sector employs around 300 people, producing approximately 100,000 garments per month. With 33 companies in the clothing segment, production, according to information on the website of one of the manufacturers, serves a large part of the national territory, being present in seventeen states of the federation.

In Jataí, there is also a direct link to points in a higher network. This can be seen in the presence of multinationals such as Louis Dreyfus Commodities and NESTLÉ, as well as the recent presence of Perdigão, which acquired a poultry factory in the town[7] . The text on the Louis Dreyfus Commodities website shows the strategic importance of Jataí in the aforementioned network:

> Present in most of Brazil's soybean-producing regions, LDCommodities owns and operates 48 warehouses, a network of port and storage facilities and three crushing plants, strategically located in Ponta Grossa, in the State of Paraná; Jataí in the State of Goiás and Alto Araguaia in the

[7]This is Frango Gale - "The company, founded in 1993, initially under the name Frango Gale Frigorífico de Aves Ltda, was founded by seven businessmen from Jatai, most of whom were farmers" (SEPLAN, 2005, p.63).

> State of Mato Grosso.[...] (www.ldcommodities.com, emphasis added).

It is also worth mentioning tourism, which is emerging as a new area for investment:

> Sites and attractions with tourism potential have been catalogued and mapped. These include museums, lakes, hot springs, waterfalls, rivers, rapids, old farms, geological formations, in short, a wide range of natural and historical resources that increase the potential for investment in tourism (SEPLAN, 2005, p.64).

These aspects demonstrate a "Zone A Centre" (IBGE, 2007b), which often goes beyond the expected levels of influence, which is reflected in the local economy, generating ever greater dynamics for the urban core.

These variables place Jataí among the most competitive municipalities in the state.

> The Competitiveness Ranking indicates the municipalities with the best conditions for attracting investment, and is therefore an important document for municipal administrators, private entrepreneurs, planners, academics and researchers, as it points out the dynamics of local economies, highlighting potential and opportunities, as well as contributing to the formulation of public development policies (SEPLAN, 2005, p.45).

These aspects also lend a certain lustre to this part of the Goiás network. In the aspects used for the ranking[8] Jataí obtained the following results: Dynamism - 6th place; Economic Wealth - 7º; Economic Infrastructure - 8º; Quality of Life - 9º; Labour - 7º; Technological Infrastructure - 6º; Financial and Tax Incentive Policies - 10th. It thus came 8th in the ranking.

In comparative terms, Jataí is behind Rio Verde in all items and ahead of Mineiros in most, with the exception of quality of life. It is therefore an important urban centre in the region, exerting an influence on other municipalities as reported.

From this perspective, Jataí reveals transformations in its urban

[8]For more details on the classification see SEPLAN, 2005.

setting, which refers to the aspect analysed by Soares (2007), who highlights the interactions and articulations that redesign part of the Brazilian territory as a result of the installation of large industrial or commercial enterprises in small and medium-sized centres.

Although Jataí has less than 100,000 inhabitants and is therefore classified as a small city according to the IBGE, its influence in the region allows it to be considered a medium-sized city in the context of the state of Goiás.

> This is because [in agreement with Deus (2002)] what defines a medium-sized city is its function, its degree of polarisation, its services and leisure facilities and the role that its urban structure plays in the region, receiving and emitting externalities. In other words, a medium-sized city is nothing more than a city with a population above the regional average, which exerts an influence on a particular sub-region, with functions that make it assume the role of regional hub in the urban hierarchy, providing the productive and collective consumption of the sub-region where it is located (DEUS, 2002a, p.244).

Since medium-sized cities serve as points of service on a regional scale, varying in size and area of operation according to regional characteristics (BRANCO, 2007), it is essential to incorporate regional functions into the concept of city, reducing the distortions produced by classifications based on population numbers in a country of continental dimensions such as Brazil (DEUS, 2002a).

Soares (2007), based on data from UNESCO (1999), highlights some of the characteristics of medium-sized cities:

> 1. are centres that provide more or less specialised goods and services for the population of the municipality itself and others, more or less nearby, over which it has influence;
> 2. are centres of social, economic and cultural interaction;
> 3. are settlements interconnected by infrastructure networks that connect to local, regional and national networks and even, in some cases, with easy access to international networks;
> 4. are centres that house local, regional and sub-national government administration bodies;
> 5. are centres that are easier to govern and control, and which in principle allow for greater popular participation

in the government and management of the city;

6. are centres that allow for greater human sociability. Places where it is relatively easier for citizens to create an identity with the place and the city itself;
7. does not have the same environmental problems as megacities;
8. Due to its size, it has less social and cultural diversity;
9. also have a lower capacity for economic competitiveness compared to metropolises or large cities, which tend to concentrate the system's higher functions. Thus, medium-sized cities have greater difficulty in accessing the main flows of information and capital.

(SOARES, 2007, p.464)

In Jataí, we can see the characteristics raised by Soares (2007), and, more specifically, the functions described by Deus (2002a) for the state of Goiás, and by Melo (2003), highlighting the economic changes produced by modern agricultural production, which are expressed spatially in the city:

> As a concentration of economic activities, the urban space has within it the potential of regional activities. In the state of Goiás, due to the strength of agricultural production, the cities have the function of ensuring production in the countryside, providing food, goods, technical assistance, etc. for this production, regardless of the number of inhabitants [...] As the modernisation of agriculture and the rural exodus progressed, the socio-economic structure of these centres began to diversify. They became regional administrative centres, with civil servants from all three spheres of power. Trade and services diversified to meet growing demand, fuelled by technical innovations. This new economic organisation led to the emergence of small and medium-sized industries, which required an increase in infrastructure and labour training, paving the way for these places to be home to medium and large-sized industries in the 1990s to transform the raw materials produced by farming into products to be consumed throughout the country and abroad (DEUS, 2002a, p.235-239).

Corrêa (2007), in a debate on a possible typology for Brazilian medium-sized cities, proposes three types of medium-sized city: a central place, characterised by a powerful concentration of goods and services for a regional hinterland; a centre of specialised activities; and a third, which is closer to the reality in Jatai:

> [A] centre for the drainage and consumption of land income. This is a city located in a traditional pastoral area, characterised by large rural properties and the absenteeism of their owners, who live in the city. This type of medium-sized town also emerges in areas where an agro-industrial complex has been set up. In both cases, the city has a significant concentration of retail and service activities, but their main clientele is the landowning elite, for whom luxury shops, restaurants, clubs and sophisticated services have their fundamental reason for existing. These activities are primarily non-basic activities and, secondarily, central functions. Through land ownership, this type of medium-sized city economically and politically controls an important regional space, the space in which its landowning elite operates. (CORRÊA, 2007, p.31)

The city of Jataí is therefore considered to be a medium-sized city in its region, in agreement with Branco (2007), who, in an attempt to identify centres that play, or have the possibility of playing, an intermediary role between the different scales of centres in the urban network, presents three groups[9] of medium-sized/intermediate cities in Brazil. Goiás includes the cities of Anápolis, in the second group, and Itumbiara, Jataí, Rio Verde and Catalão, in the third group.

More than simply understanding the size of the city of Jataí, this understanding is fundamental for the proposed analyses, in the context of an urbanisation whose urban growth has reproduced contradictions in medium-sized cities that are difficult to resolve (CARVALHO, 2003). The data presented here allows for some considerations about the implications for urban space resulting from the role played by this urban centre in the region (DEUS, 2002a). At first, they reveal the consolidation of a basically urban municipality. Marked by the small percentage of the population living in rural areas, Jataí's urbanisation rate varied by 5.85% between 1991 and 2000. The figures change from 85.36% to 91.21% respectively, compared

[9]The definition of the groups took into account, among other aspects, population and economic size, quality of life, administrative centrality and relationship life. For more details see Branco (2007).

to 87.88% urbanisation in the state of Goiás (Table 10). Despite standing out for its economic dynamism in the region, 9.41% of the inhabitants depend on government transfers to support themselves. With an average per capita income of R$304.90, which is above the state average, the injustice present in its distribution is masked: only 20% of the population hold 64% of the municipality's income. Although this rate decreased between 1991 and 2000, it is still worrying. While the change in population over the same period is 22.57 per cent, the drop in income concentration is minimal: just 1.4 per cent. In terms of access to basic services, more than 90 per cent of the population is served by treated water and electricity. This percentage drops to just over 80% when it comes to rubbish collection.

Table 10 - Municipal and State Profile - 1991/2000

Theme	Jataí		Goiás	
	1991	2000	1991	2000
Urbanisation rate (%)	85,36	91,21	80,81	87,88
Mortality up to 1 year of age (per 1000 live births)	25,6	17,8	29,5	22,5
Illiteracy rate (%)				
7-14 years	15,8	5,5	18	6,1
15-17 years	5,4	1,3	5,9	1,7
18-24 years old	7,6	2,6	7,2	2,9
Average per capita income (R$ of 2000)	236,3	304,9	211,9	286
Proportion of Poor (%)	31,2	20,3	35,1	26,6
Gini index	0,6	0,59	0,58	0,61
Percentage of Appropriated Income by Population Extracts (%)				
Poorest 20%	3,3	3,2	3	2,5
40% poorer	9,6	9,9	9,4	8,6
60% poorer	19,1	20	19,6	18,3
80% poorer	34,6	36	36,8	34,4
Richest 20%	65,4	64	63,2	65,6
Access to Basic Services (%)				
Piped water	73,8	90,4	70,6	88,7
Electricity	88,4	98,9	87,2	97,2
Rubbish collection	61,8	83,8	64,7	91,3
Access to Consumer Goods (%)				
Fridge	67,4	88,9	64,6	86,9
Television	73,3	91,5	69,7	89,1
Telephone	13,7	37,9	13,7	38,5
Computer	*	6,6	*	6,4
Human Development Index				
Municipal	0,716	0,793	0,700	0,776
Education	0,775	0,874	0,765	0,866
Longevity	0,687	0,776	0,668	0,745
Income	0,685	0,728	0,667	0,717

*Not available

Source: UNDP/IPEA/FUNDAÇÃO JOÃO PINHEIRO, 2003; AGÊNCIA AMBIENTAL DE GOIÁS, 2003. Data organisation: Márcio Rodrigues Silva.

The urban population is growing and the problems are concentrated in the city, since the population tends to migrate there. An example of this is the proportion of poor people in the municipality. The rate fell by 10.9 percentage points over the period in question, dropping from 31.20 per cent to 20.30 per cent. However, the absolute number of people in this situation grew from around 13,900 in 1991 to 15,300 in 2000, more than the population added to the municipality.

With regard to urban voids in 2004, according to data from the local town hall (JATAI-GO, 2004b), the city of Jataí had a total of approximately 37,000 urban plots, of which around 13,000 were unoccupied, i.e. 35 per cent of the total.

Could these problems be the price to pay for the city's growing influence in the region? We don't think so! The challenge is to *develop,* in its authentic sense, and not just to *grow.*

2.2 - CHARACTERISING THE URBAN FABRIC: IDENTIFYING THE FORM IN ORDER TO ANALYSE ITS CONTENT

After the previous description, the researcher's gaze requires a closer look at the object of study. Increasing the scale of the work, from a single point in the urban network to a detailed patchwork quilt, also means expanding the space for enquiry and discussion in order to build a scientifically based understanding of the urban area of Jataí. This exercise was based on the following data sources:

- Real Estate Registry of the Municipal Finance Department for 2003 (JATAÍ-GO, 2003), used to compile data on urban voids;
- Programme for the registration of economic activities by the Secretariat for Fiscal Management in 2004 (JATAÍ-GO, 2004c), which provided the economic data;
- A report on the economic activities registered up to 2004 (JATAÍ-GO, 2004d), from the Fiscal Management Department, provides an overview of the economic activities set up in the city;
- Urban Master Plan for the years 2001 and 2003 (JATAÍ-GO, 2001 and 2003), making it possible to verify the municipality's urban policies;
- Preliminary survey of vacant lots in 1995 (JATAÍ-GO, 1995) serving as a parameter for a comparative analysis of urban voids; and
- Cadastral maps for 2003 and 2004 (JATAÍ-GO, 2003a and 2004), supplied on digital media in DWG format.

Once we had this data, we proceeded to create the map base[10] , necessary for the proposed analyses, by converting the data in DWG format to SHP (Shapefile) format. As the original file was disorganised, showing the polygons for the city's neighbourhoods in the same layer and grouped together, the decision was made to redraw the neighbourhoods based on the original file, which had already been georeferenced.

To do this, the data from the cadastral maps was added and a Shapefile called "neighbourhoods" was created in the ArcCatalog

[10]ArcGis® software was used

module. Then, using the "Editor" tool, the polygons for each neighbourhood were digitised and the following data was added (Figure 6): *!d* - number assigned to identify the neighbourhoods; *Name* - Name of the neighbourhood and *Dt_impl*- Date of implementation of the neighbourhood.

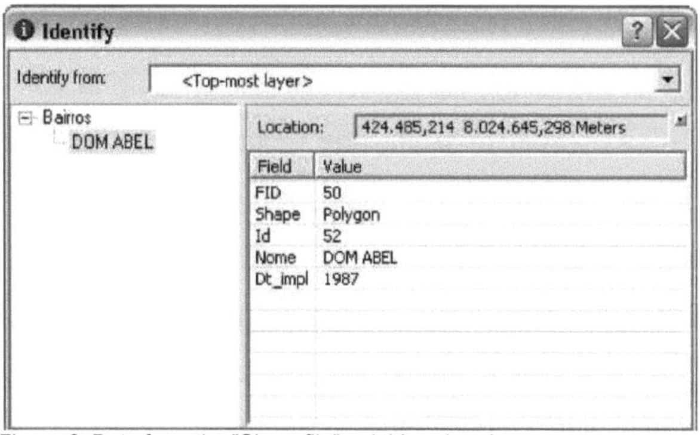

Figure 6: Data from the "Shapefile" neighbourhoods

The basic table was joined with data from existing tables in spreadsheet format, which were converted to database format (DBF). refer to the percentage of vacant plots (1995 and 2004) and the number of registered establishments per decade, in the categories of commerce, industry and service provision. After the "join" operation, the following screen resulted (Figure 7).

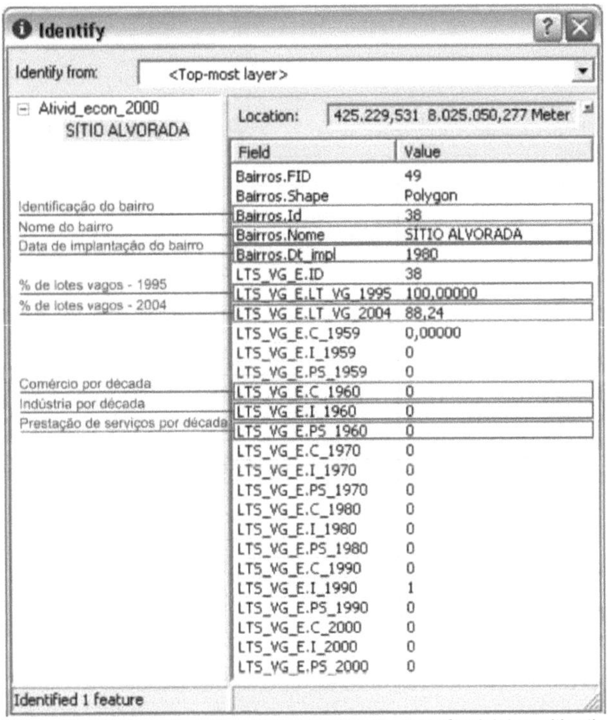

Figure 7: Data from the "Shapefile" neighbourhoods after joining with other tables

2.2.1 - URBAN SPRAWL MAP

The urban sprawl map (Figure 9) was generated using the "Date the neighbourhood was established" field via the "symbology>quantities>graduated colours" tool (Figure 8). Classification by decade was chosen to make it easier to analyse and cross-reference the data with other variables available only every ten years. The use of ordered colours allowed for a visual assessment of urban evolution through the implantation of neighbourhoods.

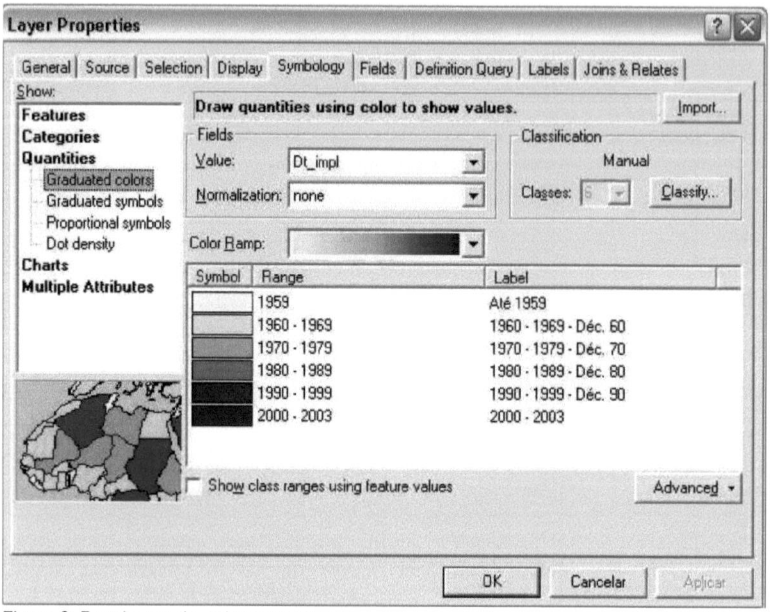

Figure 8: Drawing up the urban sprawl map

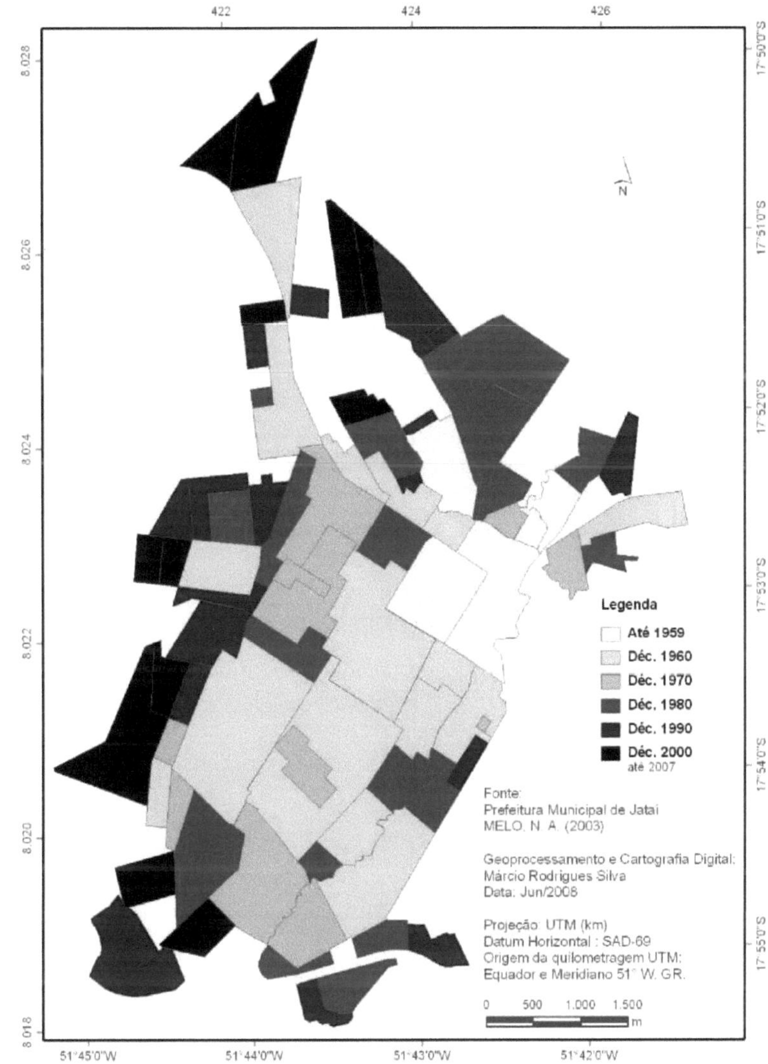

Figure 9: Jataí-GO - Urban Expansion: Allotment Developments

2.2.2 - MAP OF URBAN VOIDS

To make the map of urban voids (Figures 11 and 12), the "% of vacant plots" field was used for 1995 and 2004, applying the "symbology>quantities>graduated colours" module (Figure 10). The variables were classified into four ranges corresponding to "25%

vacant lots" each, in order to simplify the analysis and standardise the 2004 data with that of 1995, the latter being available only in these ranges. The use of ordered colours was aimed at visually reading the high percentage of vacant lots in the city.

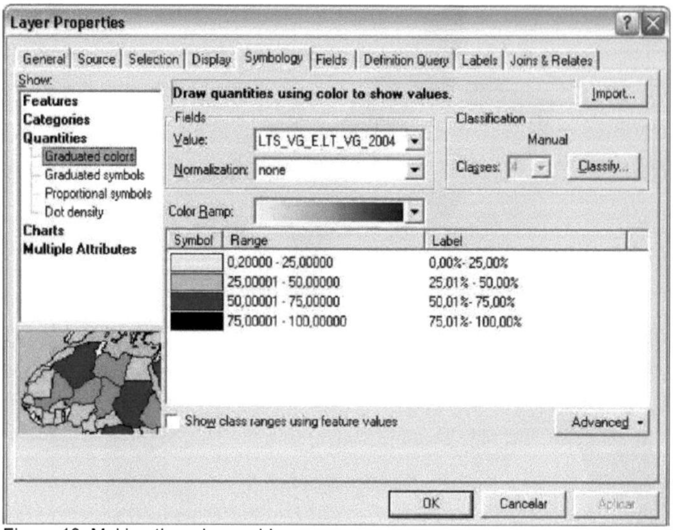

Figure 10: Making the urban voids map

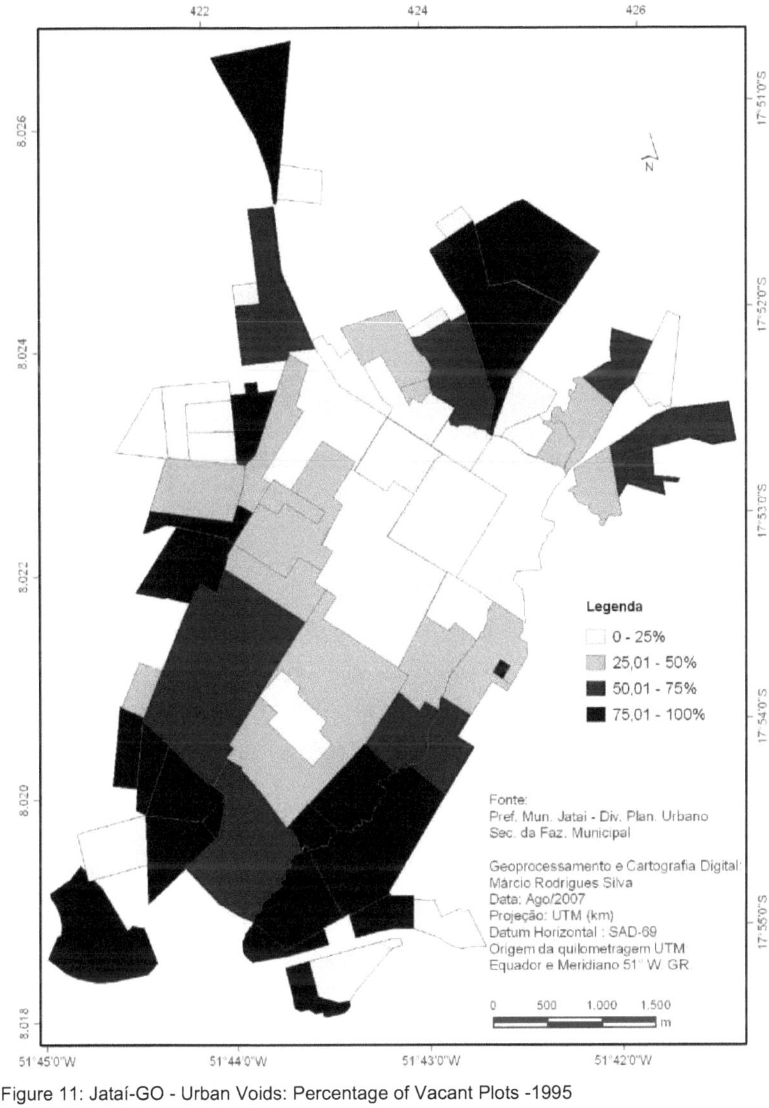

Figure 11: Jataí-GO - Urban Voids: Percentage of Vacant Plots -1995

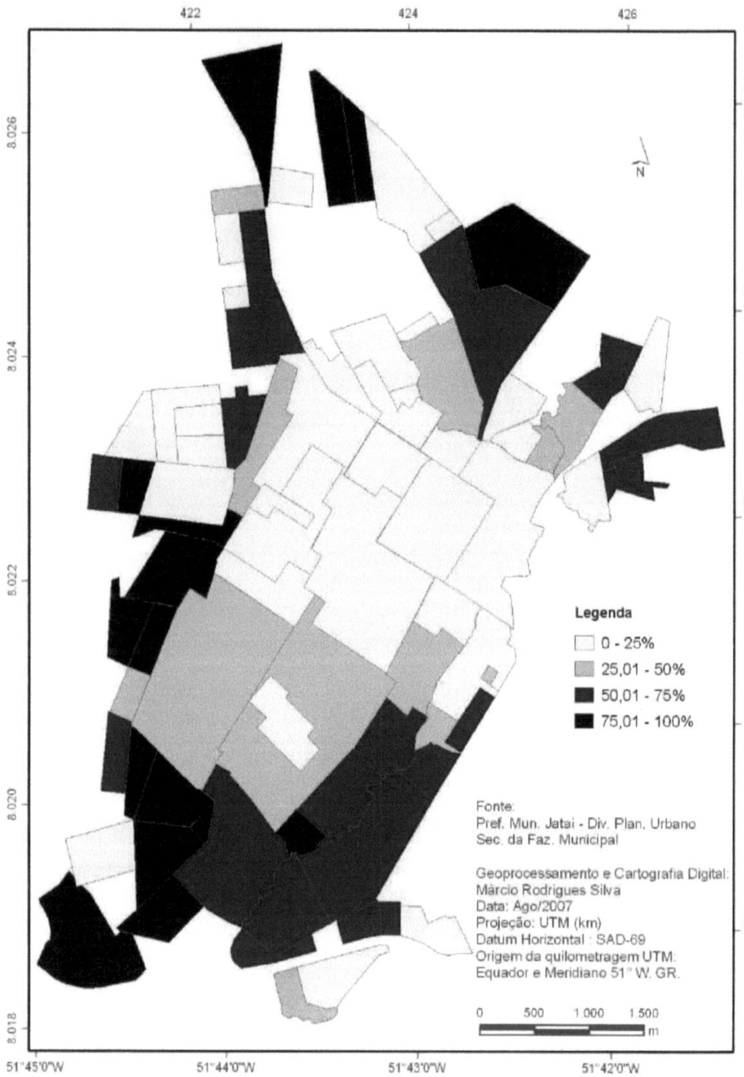

Figure 12: Jataí-GO - Urban Voids: Percentage of Vacant Lots - 2004

2.2.3 - MAPS OF ECONOMIC ACTIVITIES

The maps referring to economic activities in the city of Jataí (Figures 15 to 20) were drawn up using the "symbology>quantities>pie" tool (Figure 13). The fields used were "trade, industry and provision of services by decade". The result was a collection of maps that made it possible to read the evolution of the economic activities implemented in each selected decade. To do this, the variables corresponding to the decade analysed were selected for each map and the size of the symbol was classified according to the sum of the values used (Figure 14). The minimum size was standardised at "20 units" to take account of the variation in values across all the decades and to enable a visual comparison between the maps.

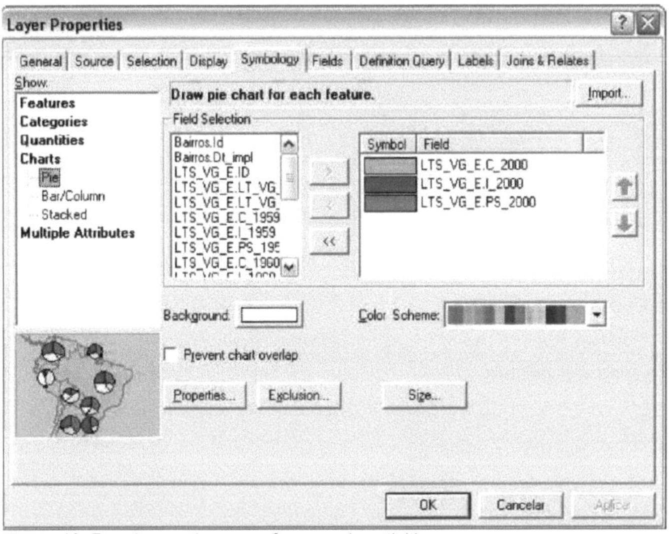

Figure 13: Drawing up the map of economic activities

The use of selective colours, with one colour for each economic activity, aimed to show the proportion of each activity installed in the neighbourhood. The diameter of the circle reflects the sum of the activities installed in each neighbourhood in the selected decade.

The maps reflect the agglomeration of the activities mentioned in each neighbourhood, in the time frame from "up to 1959" to "2000s". The wide range between the smallest values ("up to 1959") and the largest ("2000s") consequently resulted in very distinct maps in which the main focus was to make it possible to identify the agglomeration of the variables being worked on.

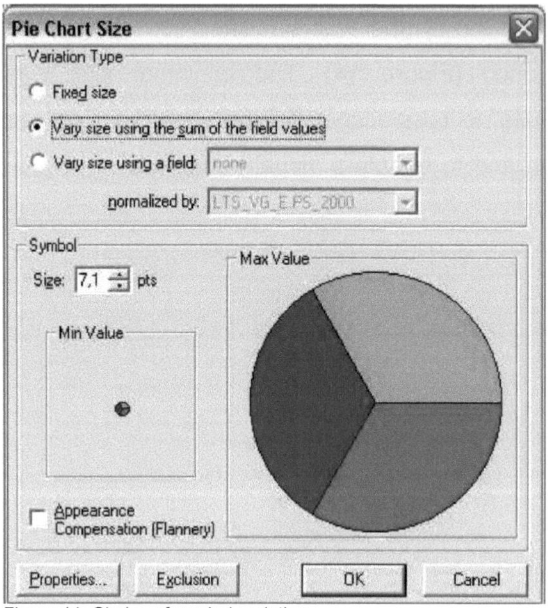

Figure 14: Choice of symbol variation

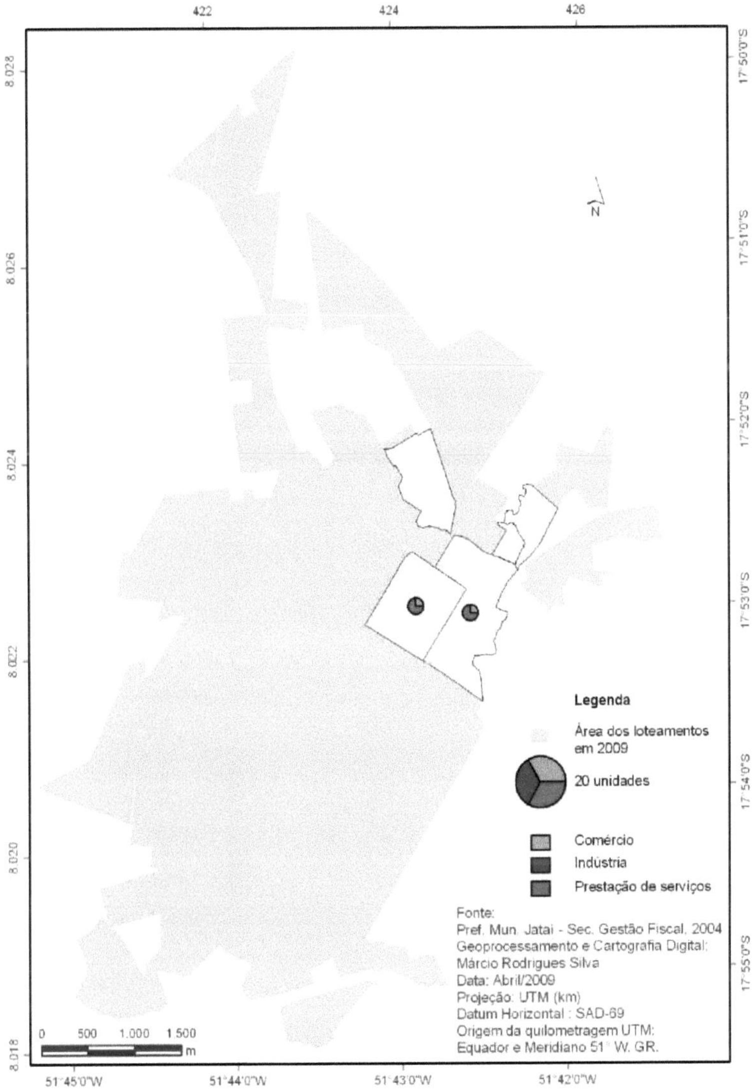

Figure 15: Jataí-GO: Economic activities set up until the 1959s

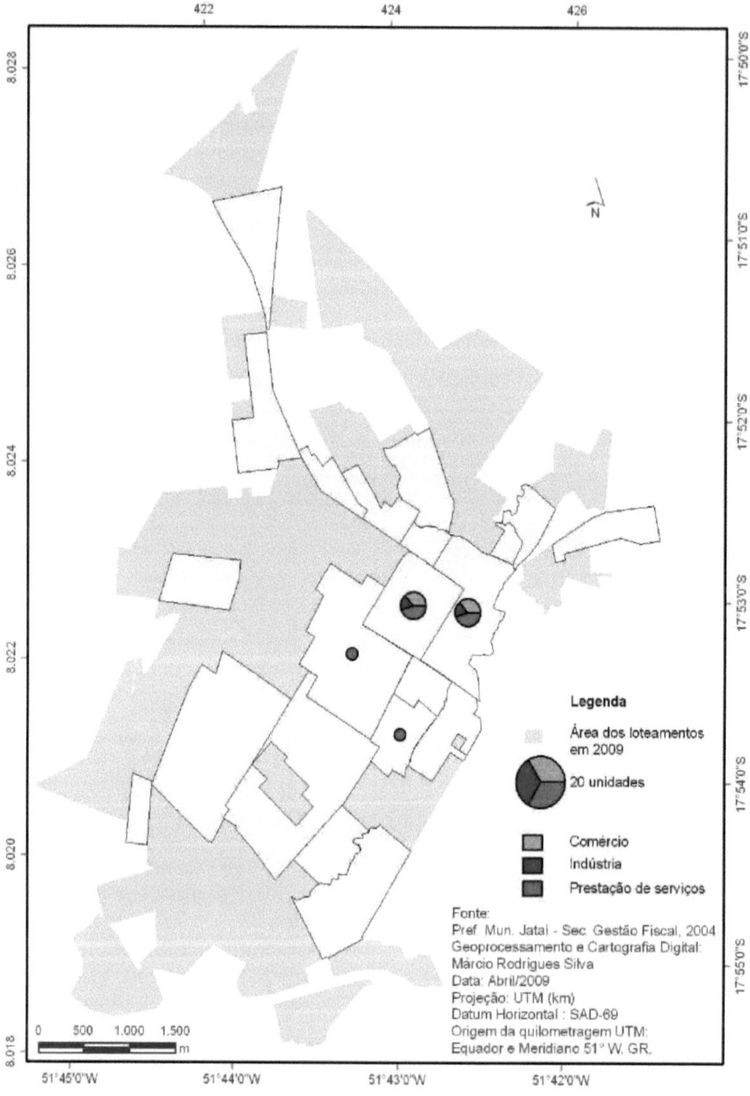

Figura 16: Jataí-GO: Economic activities set up in the 1960s

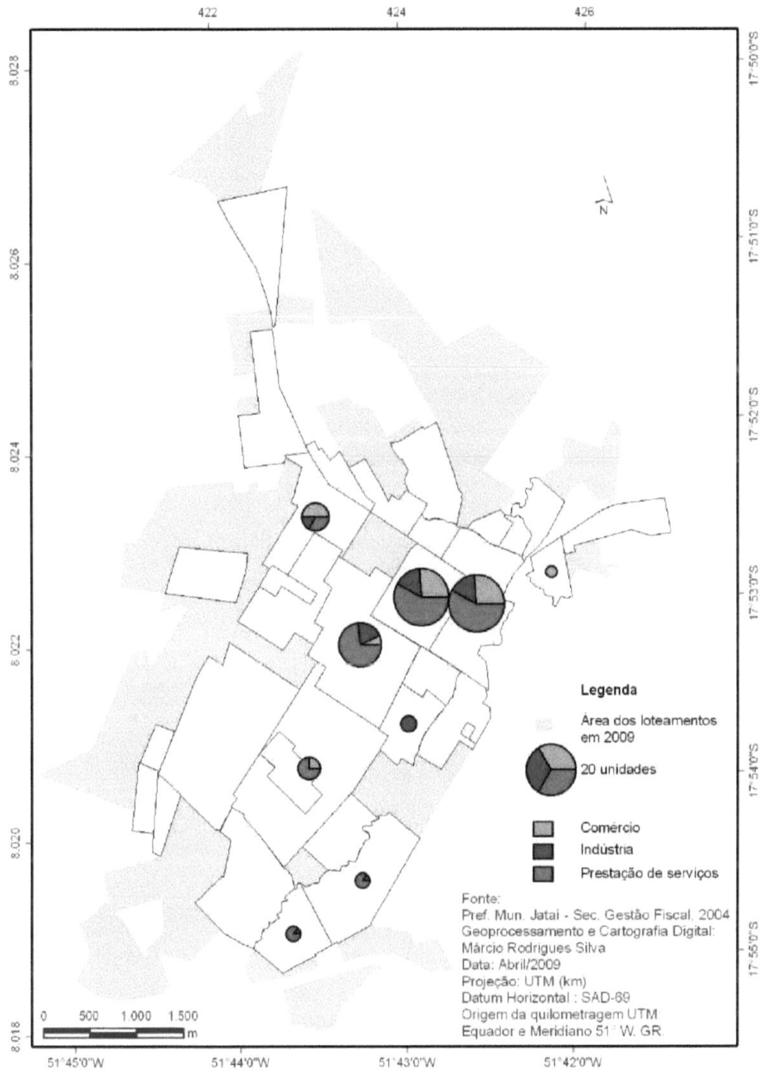

Figura 17: Jataí-GO: Economic activities set up in the 1970s

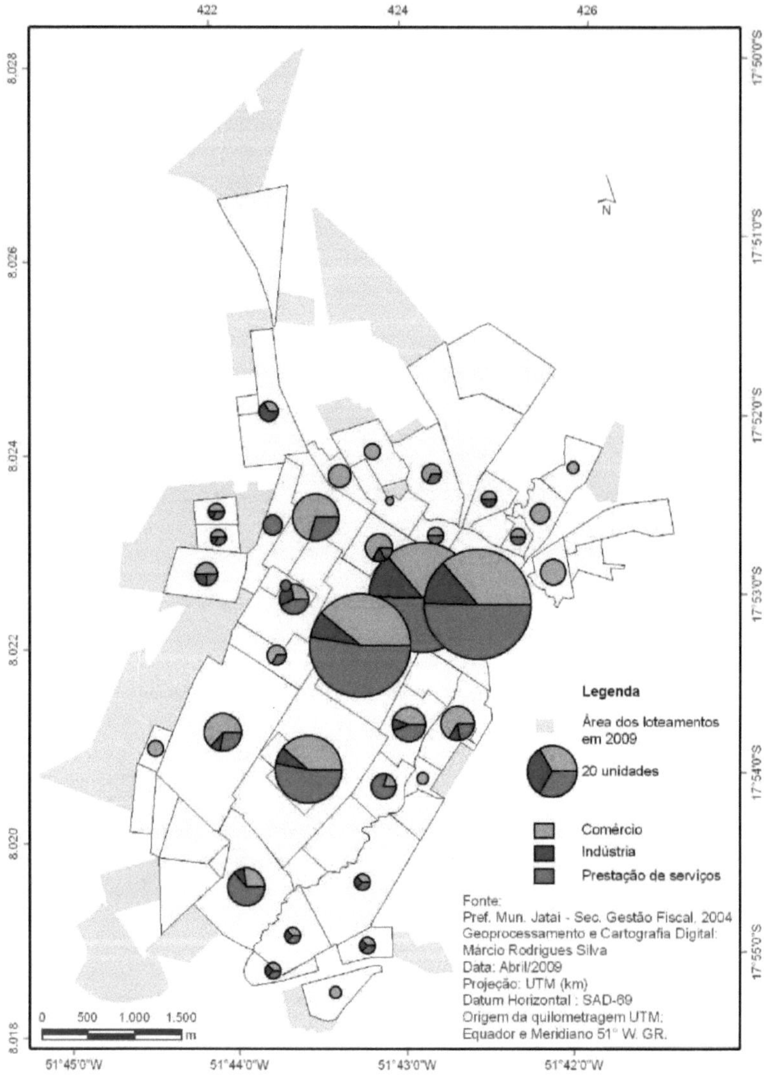

Figura 18: Jataí-GO: Economic activities set up in the 1980s

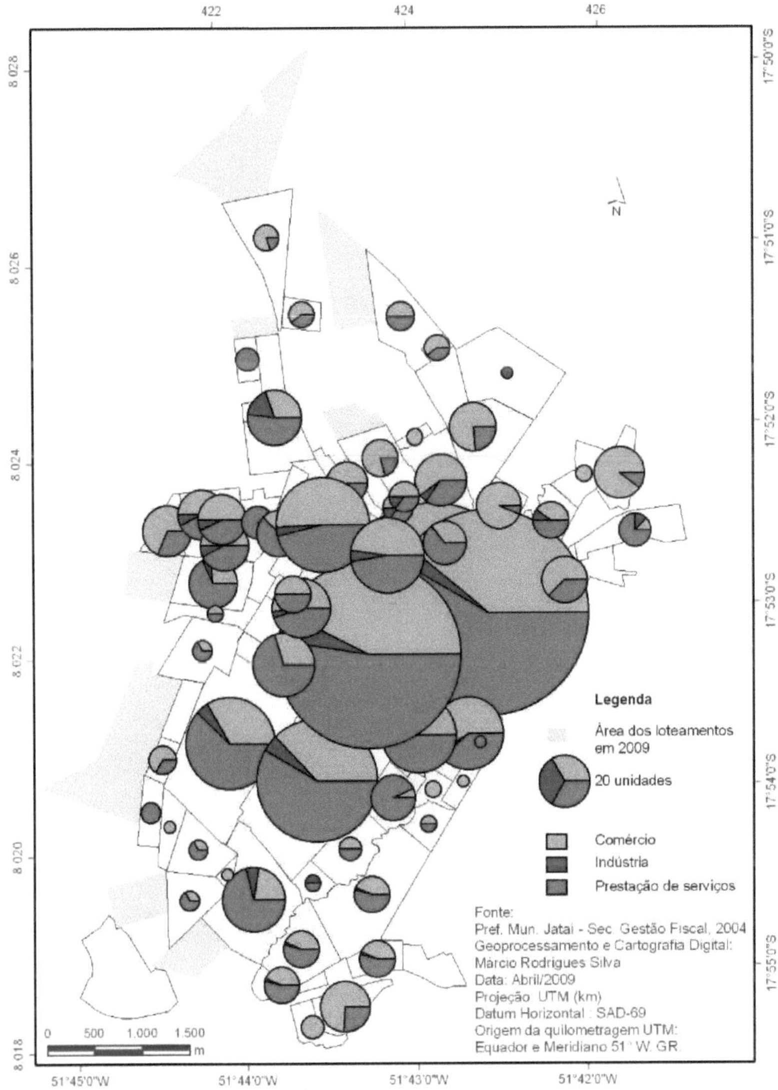

Figura 19: Jataí-GO: Economic activities set up in the 1990s

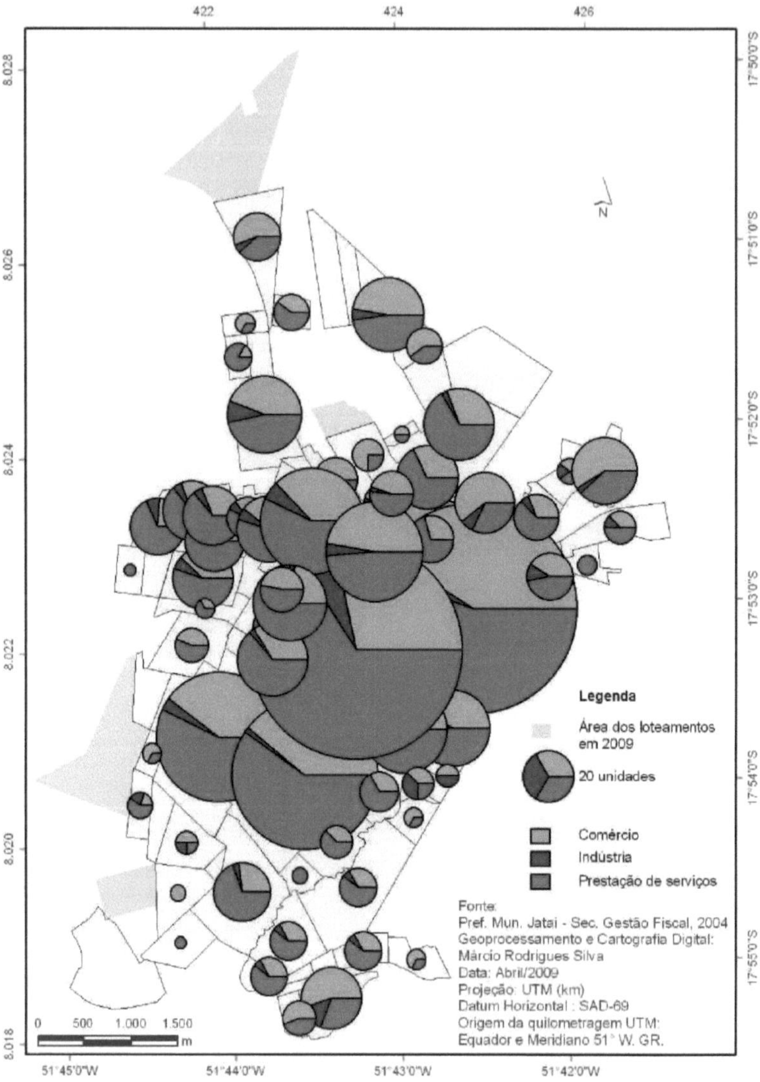

Figura 20: Jataí-GO: Economic activities set up in the 2000s

2.2.4 - ECONOMIC ACTIVITY CONCENTRATION MAP - DYNAMIC AREAS

The map of the concentration of economic activities (Figure 22) seeks to summarise the previous maps by the concentration of economic activities, by decade, brought together in the same representation. The sum of the "trade, industry and service provision by decade" data was used, represented by the "symbology>quantities>dot density" tool (Figure 21). The use of dots corresponding to five units each, on an ordered colour scale, made it possible to analyse the number and concentration of establishments set up per decade in each neighbourhood and in the urban area as a whole. In this way, the city's dynamic areas were defined, as analysed below.

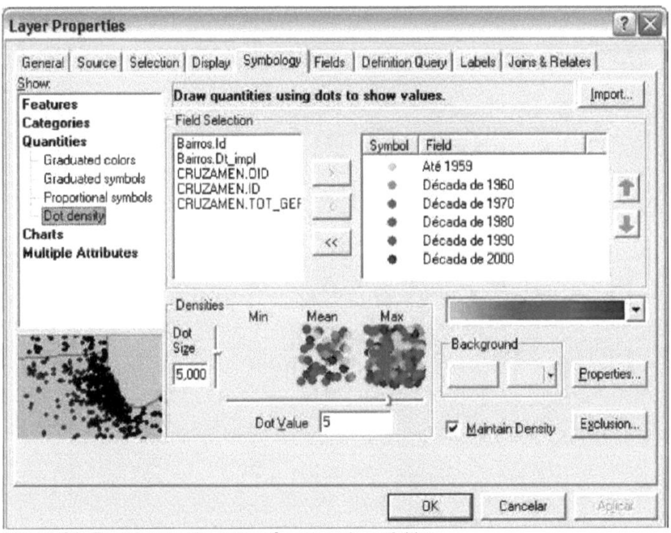

Figure 21: Drawing up the map of economic activities

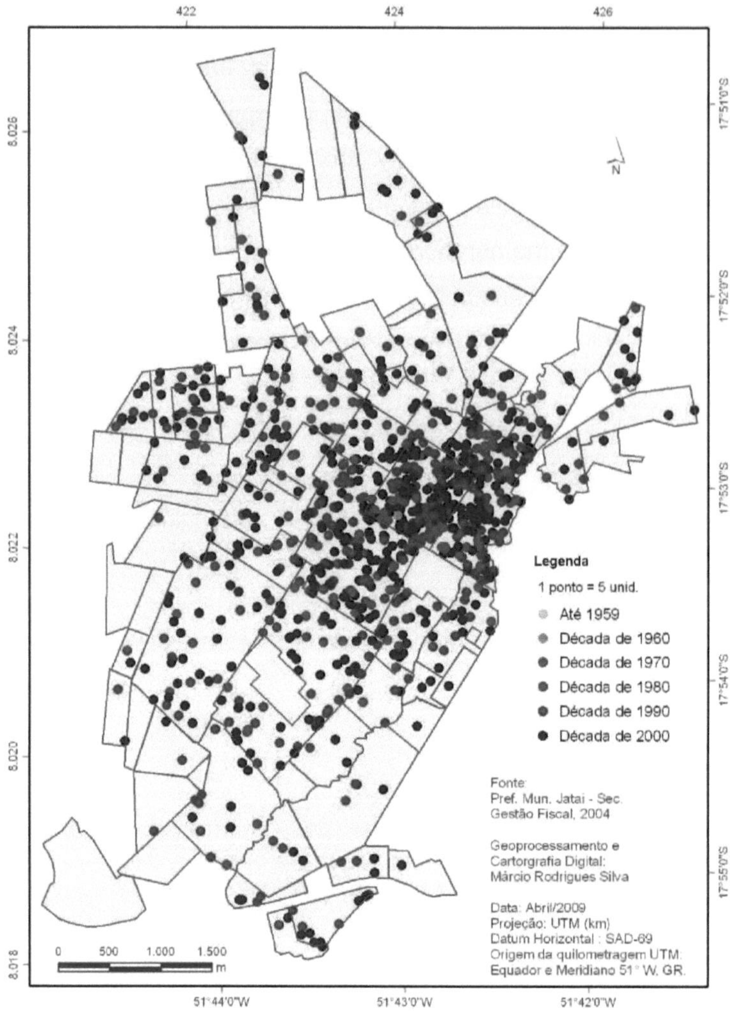

Figure 22: Jataí-GO: Dynamic areas: Concentration of economic activities

2.2.5 - REFLECTIONS ON CHARACTERISING THE URBAN FABRIC

The urban space found in Jataí today is the result of construction that began in the second half of the 19th century, when its first occupants settled here. These were to become the residents of Jataí's first neighbourhoods. The city's original centre, initially located on the banks of the Jataí stream, would expand in approximately 120 years, if we consider the year 1836, when this occupation began, reaching an area of 1.73 km^2 (Figure 23). In subsequent decades, urban expansion took place at a new pace.

The area occupied by allotments in the 1960s was 10.46 $kilometres^2$, an increase of more than 500% on the 1959 figures. However, the allotted area is not actually occupied. Analysing an aerial photograph from 1965, we can see that the area actually occupied, implying buildings and not just open plots, is 6.13 km^2 (Figure 24). This figure corresponds to 59% of the allotted area. Thus, even at the beginning of urban development, the vision of valorising urban land by setting up allotments on the outskirts of the city with a view to valorising the intermediate areas is perceptible.

In the 1970s, Jataí's urban area grew by 28% compared to the previous decade, reaching a value of 13.41 km^2. From 1970 to 1980 there was a 40% variation, with the occupied area totalling 18.88 km^2. In the 1990s, the urban area totalled 22.14 km^2, 17% more than in the previous decade. Currently, the occupied area is 25.60 km^2, reflecting a 15% increase over the 1990s.

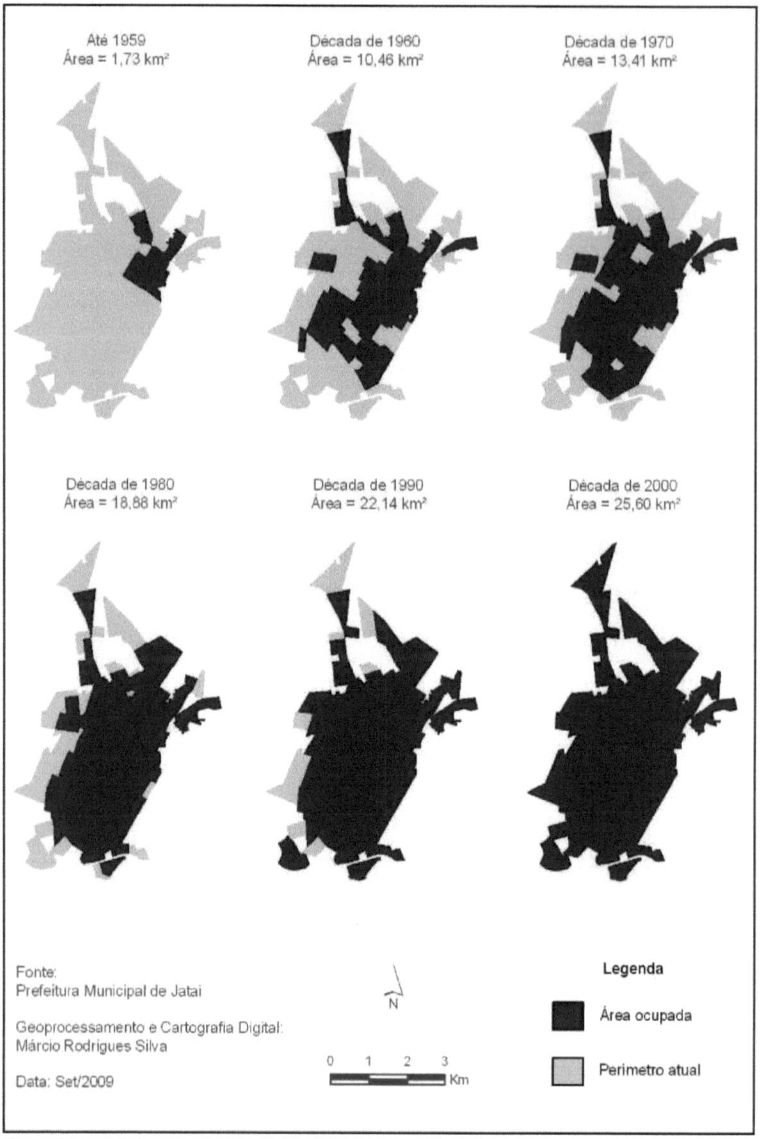

Figure 23: Jataí-GO: Implementation of allotments - Evolution of the occupied area

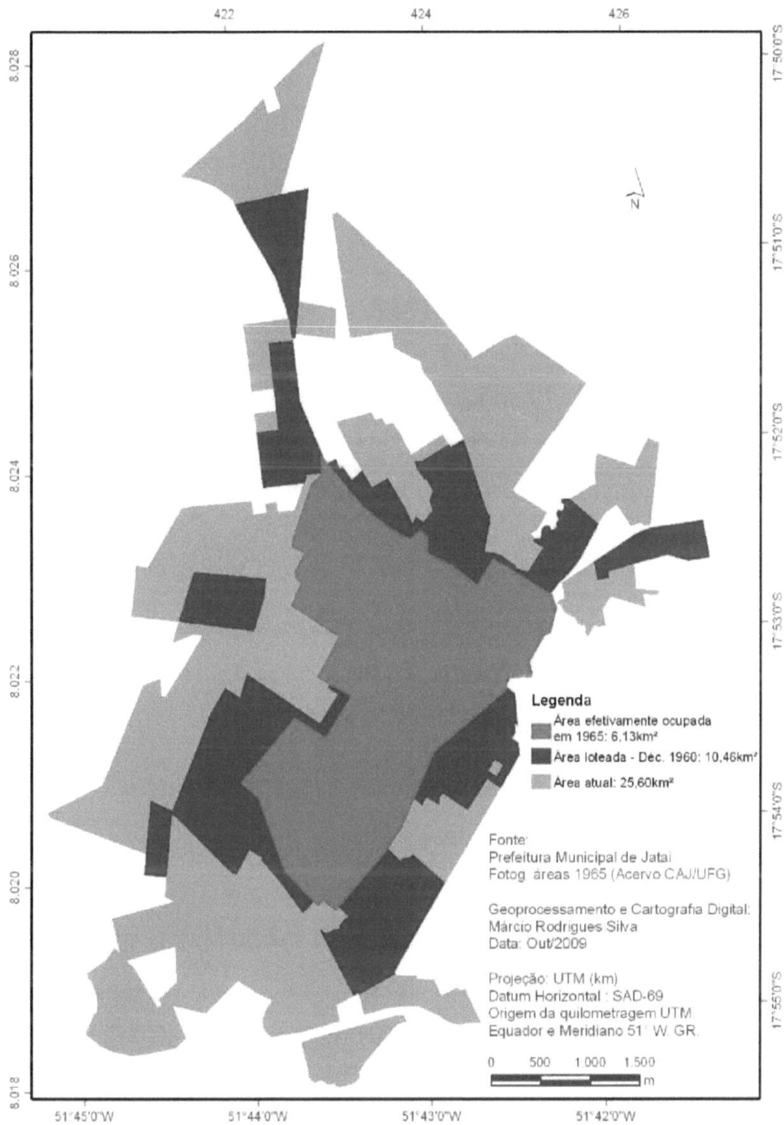

Figure 24: Jataí-GO: Area effectively occupied in 1965

The process of capitalist reproduction imposes its actions without considering that the city also houses the individual, who has basic needs such as food and housing. Lefebvre (1999, p.25), commenting on Engels' view of the situation of the working class in

England, says, based on Engels' words, that it is much easier to blame the city, or general immorality, or the forces of evil, than to tackle the problem at its true level: politics. Politics here doesn't just refer to local politics, past or present, but above all to the oppressive capitalist action that excludes a large section of the population from access to the most basic human goods.

The imposition of capitalism, which subjects individuals to its rules, is one of the factors that generates changes in spatial configuration. Table 11 shows the significant presence of low-income housing estates in the 1980s and 1990s in the city of Jataí, an aspect that extends, at least at first, into the current decade.

Table 11 - Jataí-GO: Allotments by typology

Decade	Until 1959	1960	1970	1980	1990	2000	Total
Grand Total of Allotments Implemented	5	19	11	22	18	12	87
Housing Estates Financed			1	3	3	3	10
Low-Income Housing Estates Low Income				2	7	2	11

Source: Prefeitura Municipal de Jataí; MELO, N. A. (2003). Data organisation: Márcio Rodrigues Silva. Data for the 2000s refers to the period 2000-2008.

The emergence of such subdivisions is an "urban response" to the changes imposed on Jataí.

Financed housing developments began in the 1970s. Low-income housing developments began in the following decade. An important fact to consider is that the 1980s were marked by the expansion of the agricultural frontier in the municipality, and it was in this decade that the demand for affordable housing for the low-income population arose.

The growth of the local economy also led to a dynamic property market for the middle classes. In the 1980s we had the largest number of allotments set up, with three financed housing developments standing out.

The same decade also saw the emergence of two low-income housing developments, reflecting the demand for housing that could not be met through financial acquisition, necessitating "assistance" from the state.

The 1990s saw a continuation of the previous period with a total of eighteen housing developments, three of which were financed and seven for the low-income population. The high number of the latter type of allotment raises several debates, including social inequality, which leads to a demand for low-income housing that is three times greater than in the previous decade.

The current decade suggests growth at the same rate as the previous two, since by 2008 a total of twelve allotments had been registered, three of which were financed and two of which were low-income.

Another fact to consider when analysing Jatai's urban space is the occupancy rate of the allotments set up. A comparison of data provided by the local town hall for 1995 and 2004 shows that although the occupancy rate has increased in general terms, the city's vacancy rate is still significant (Table 12).

Table 12 - Jataí-GO: Percentage of vacant plots in relation to the total allotted area - 1995 and 2004

	% of voids per strip		
Void range (%)	1995	2004	Variation
0 a 25	35%	35%	0%
25,01 a 50	23%	20%	-3%
50,01 a 75	17%	23%	6%
75,01 a 100	25%	22%	-3%
Total	100%	100%	

Source: Jataí City Hall; Data Organisation: Márcio Rodrigues Silva.

Table 12 (see also figures 11 and 12) shows that in 1995 25% of allotments had between 75.01% and 100% of their plots unoccupied, whilc in 2004 22% of allotments had this rate. In this range, which is considered the most worrying as there are totally unoccupied allotments, there has been a reduction of 3%, a percentage that could be increased if new allotments were not set up before the vacant areas were occupied.

In the 50.01% to 75% range, there was an increase of around 6% from 1995 to 2004, due to the reduction in the critical range and the implementation of new housing developments.

In the range of 25.01 to 50% of vacant plots there is a drop of 3%, demonstrating the occupation of some intermediate areas (between the centre and the periphery).

In the ideal range for allotments, with total occupation, or a vacancy rate of no more than 25% of plots, the variation is zero, due to the increase in the number of neighbourhoods. In this range, there is a significant evolution masked by new neighbourhoods. The tendency is for intermediate areas to be occupied as new neighbourhoods emerge. Although the prices for acquiring these areas are high, there is a demand for them, leading to their occupation.

Figures 11 and 12 show a spatialised view of this data. There is greater occupation of the peripheral areas of the city and of areas that were valued between 1995 and 2004. The large occupation of some peripheral areas is due to the presence of housing estates that also play the role of adding value to more central areas, since some infrastructure, which is not ideal, is made available to supply these locations. This shows the state as a driver of urban occupation.

After explaining the process of urban expansion in Jataí, one question remains: Is there a need for urban expansion in terms of adding new areas?

The occupation of existing areas, before the opening up of others, would bring some benefits to the urban complex in question:

The neighbourhoods established in more recent times are located on the outskirts of the urban fabric. Of the 30 neighbourhoods set up between 1990 and 2008, 15 are densely populated areas, these being housing estates and low-income housing estates (Table 13);

- Bearing in mind that a large proportion of the residents of these localities work in the more central areas, where the city's main economic activities are located (Figures 15 to 20), transport would be made easier as the distance between home and work would be shorter;

- In terms of urban infrastructure, such as transport, the public authorities could concentrate their actions more effectively. The area served would be smaller and the population reach greater. Likewise, important aspects of the population's quality of life would see a significant improvement in their conditions. Health, education, leisure, among others, would be enhanced in a concentrated development prioritising the quality of services.

Table 13 - Jataí-GO: Implementation of Allotments

Decade	Sector	Year of Implementation	Total
Until 1959	Neutral Field	until 1959	5
	Centre	until 1959	
	Centre (lower part)	until 1959	
	Santa Terezinha	until 1959	
	Vila Palmeiras	until 1959	
1960s	Brasilia	1.960	19
	Fatima 1." Stage	1.960	
	Jardim Rio Claro	1.960	
	Olavo	1.960	
	Progress	1.960	
	St Lucia	1.960	
	Santa Maria	1.961	
	Iracema	1.962	
	Jardim América	1.962	
	Fatima 2.ª Stage	1.963	
	Three Marias	1.963	
	Bela Vista	1.964	
	Fatima 3.ª Stage	1.965	
	Sofia	1.965	
	Epaminondas I	1.966	
	Jardim Paraíso	1.966	
	Popular	1.966	
	Farmer	1.968	
	Bom Conselho College Area	1.968	
1970s	Oak	1.972	11
	Industrial	1.975	

	Luiza	1.975	
	Airport	1.976	
	Friar Domingos	1.976	
	St Peter's	1.976	
	Vila Paraíso	1.976	
	Sofia II	1.978	
	West	1.979	
	Samuel Graham	1.979	
	Sodré	1.979	
1980s	Antenna	1.980	
	Freedom Garden	1.980	
	Sítio Recanto Alvorada	1.980	
	Fabriny	1.981	
	Jose Estevam	1.981	
	Alvorada recreational sites	1.981	
	Holy Spirit Divide	1.982	
	Jardim Maximiano	1.982	
	St Anthony	1.982	
	Sofia III	1.982	
	Geda	1.983	
	Hills	1.983	
1980s	Alto das Rosas	1.985	22
	Vila Mutirão	1.986	
	Rio Claro Complex 1	1.987	
	Dom Abel	1.987	
	Aimbiré (Jardim Ana Paula)	1.987	
	Blue Mountain	1.987	
	Lamb	1.988	
	Plateau	1.988	
	Estrela D'alva set	1.989	
	Rio Claro II complex	1.989	
1990s	Dorival de Carvalho set	1.990	18
	Filostro Machado set	1.990	
	Rio Claro III complex	1.990	
	Conjunto Sebastião Herculano	1.990	
	Epaminondas II	1.990	
	Francisco Antônio	1.990	
	Hermosa	1.990	
	Jacutinga	1.990	
	José Bento	1.990	
	Mansions	1.991	
	Cohacol complex	1.994	
	José Ferreira (Dori)	1.994	
	Bei mar	1.994	
	St Lucia II	1.995	
	Dom Benedito	1.996	
	Beehive Park	1.997	
	Jardim Goiás Complex	1.998	
	Residencial Barcelona	1.998	
2000-2008	Spring neighbourhood	2.000	12
	Colméia Park II complex	2.000	
	Jardim Goiás II complex	2.000	
	Forest Garden	2.000	
	Cylleneo France set	2.001	
	Colméia Park III complex	2.003	
	Vila do Sol Residence	2.003	
	Cohacol V	2.005	
	Residencial das Brisas	2.008	
	Bandeirantes Residence	2.008	
	Portal do Sol 1	2.008	

	Portal do Sol II	2.008	
	Total		87

Source: Prefeitura Municipal de Jataí; MELO, N. A. (2003). Data organisation: Márcio Rodrigues Silva.

In order for this situation to be implemented, it is necessary for public authorities to take action with instruments that promote the occupation of empty urban areas, such as social policies aimed at better income distribution, giving citizens the conditions to acquire and occupy the areas in question.

Above all, existing legislation must be complied with. According to what is established in the Urban Master Plan (JATAI-GO, 2007) in its article 1°, paragraph 6°, items 4 and 4.5, there must be "planning and control of urban land in order to avoid the speculative retention of urban property, which results in its under-utilisation or non-utilisation".

Also, according to the city's Land Use and Occupation Law (JATAÍ-GO, 2008), in its chapter VIII, article number 20, the "[...] City Hall may determine the compulsory parceling, building or use of unbuilt, underused or unused urban land [...]". Other instruments, such as the application of progressive IPTU, are also provided for. According to the same law, in its article 21, in the event of non-compliance with the conditions established and the deadlines set for land occupation "[...] the municipality will apply the progressive Urban Property Tax (IPTU) over time, by increasing the rate for a period of five consecutive years".

These legal determinations follow what is established in article 182 of the Federal Constitution, in its chapter II that deals with urban policy, and in the City Statute that regulated articles 182 and 183.

2.3 - SOCIO-SPACIAL CONFIGURATION OF JATAF- GO: SOCIETY IN THE URBAN SPACE

Jataí, as a city, "[...] concentrates not only the population, but the instruments of production, capital, needs, pleasures [and] everything that makes a society a society" (LEFEBVRE, 1999, p.49) (emphasis in original).

As Santos (1998, p.10-11) notes, cities "[...] can be centres for meeting the circulation needs of agricultural production or small-scale industrial activities". In this sense, as production grows, tertiary activities diversify to meet their own internal needs. The specialised activities developed in the city, as well as those geared towards exports, "[...] generate a local dynamism that sustains a broad and varied set of activities to meet local demand" (SANTOS, 1998, p.13).

The local demand was in line with Lefebvre's (1999, p.149) statement: "Finally and above all, the city and the urban agglomeration concentrate 'services'". This dynamism can be seen in the data on the implementation of economic activities in Jataí (Table 6), which shows that service-providing establishments have been set up as a priority, as can be seen in Figures 15 to 20.

The city, being a work in itself, brings together the production of needs and pleasures that give meaning to production. It is also where goods are produced, exchanged and consumed (LEFEBVRE, 1999, p.51). All of this requires spatialisation. This spatialisation crystallises the urban space. In it we realise that the city is "[...] a social thing, in which social relations are evident (become sensible) which, taken in themselves, are not evident [...]" (LEFEBVRE, 1999, p.140) (emphasis in original).

This combination of encounters and disagreements reflects what Santos (1998, p.23) says about the creation of "agricultural regions". Their formation "[...] involves the need to establish cities of a size compatible with tertiary activities, related to the circulation [of production], as well as the provision of services that make this production viable".

The structure of local society, which is nothing more than a reflection of a broader conjuncture aimed at the reproduction of capital, is articulated with the urban phenomenon (Castells, 1983, p.43). Urbanisation reflects the relationship between society and space. It is a function of the specific organisation of modes of production, as well as the internal structure of the one that prevails (Castells, 1983, p.111).

The context of this analysis considers that economic development and urban growth are closely linked, as can be seen in the map of urban expansion (Figure 9) and the installation of economic activities (Figures 15 to 20). The 1980s and 1990s saw the largest number of allotments set up in the proposed time frame, which coincides with the most significant period in the rise of economic activity (Table 6).

Starting from the original centre, the areas to the west, with suitable terrain and no barriers to urban growth, have always been prioritised when it comes to setting up new allotments. It should be noted that the conditions in the eastern part of the city are not favourable for expansion, as it is home to the Brazilian Army's 41st Motorised Infantry Battalion, which occupies a significant area and, together with topographical conditions [11], is a barrier to growth in the eastern part.

Urban expansion, especially during the rise of mechanised agriculture, provided the necessary space for the establishment of economic activities to meet the demand driven by the dynamics of modern agriculture.

The decades of greatest local economic growth correspond to the phase of agricultural modernisation, essentially from the 1970s onwards, and its consolidation in later decades (1980s and 1990s). The period in question represents a leap in the Jataiense economy, with emphasis on service and commercial activities, respectively. Melo (2003, p. 121) highlights this change in urban space, since

> Modern agricultural production has produced economic changes that are expressed spatially in the city. At the same time, the city has changed to meet the development of production in the countryside and its internal demands for

[11] This area has steeper terrain than the city as a whole.

services and equipment.

By analysing the agglomeration of economic activities (Figure 22) in the city of Jataí, we were able to define some areas that stand out in the urban scenario. The definition of these areas, as explained above, took into account the concentration of activities installed in each neighbourhood in the decades analysed. In this way, the dynamics of an area are measured by the concentration of points corresponding to the data analysed (Figure 22). The higher the concentration of points, the more dynamic the area. These areas show a greater increase compared to the others, which have either been allotted and not occupied, or have been occupied but have no dynamics of their own, remaining opaque and dependent on other parts of the urban fabric.

The process of urban expansion was not necessarily accompanied at first by development in the regions occupied at the time. The set of factors needed to consolidate a certain sector of the city was distributed spatially, delimiting the dynamic areas (Figure 22). Starting from the city centre, the same area as its original core, there are prominent movements in the south-south-west and west-north-west directions.

The original nucleus was the centre of economic activity and the heart of the city until around the end of the 1950s. This area then expanded in a south-south-westerly direction, consolidating in the 1960s the main area of concentration for commerce and services, which is still prominent today.

In the period 1970-1980, the urban space showed a shift of some activities, previously located only in the central areas, towards the newly occupied areas.

The process of deconcentration of central activities

continued with greater force in the 1990s. During this period, the areas located to the west-north-west became more prominent on the urban scene. This decade was marked by the presence of several popular housing developments and also by the displacement of economic activities in order to cater for the population density characteristic of these locations.

In the current decade, there has been a movement towards the peripheral areas of the city with the establishment of new housing estates and housing complexes. The areas in between have been allotted land, which doesn't necessarily mean they are occupied.

Although the central region appears to be indispensable to the city's dynamism, it is clear that there are movements that tend to move the centre of decision-making to areas that offer better access conditions (mainly roads) and greater prospects for development. This movement is reminiscent of Castells' (1983) words, when he emphasises the role of the car in contributing to urban dispersion. It doesn't mean the decay of the former in relation to the latter, but rather a movement towards the gestation of a new dynamic area.

Through this analysis it is possible to get a real view of the "living city", where there is movement, presence, occupation, which differs from the legal city, determined by the official perimeter. This city reflects its society, the influences suffered in its construction process and shows an urban fabric marked by dynamic areas (Figure 22) and voids (Figures 11 and 12), which express the local urban dynamics as they shift the gaze to the city's bright spots.

The distribution of the population in Jatai's urban space is the result of a set of factors closely linked to the characteristics of local society. The process of occupying the city has brought with it reflections of diverse actions that materialise in the urban layout.

Putting the pieces of this puzzle together spatially reveals different realities: noble neighbourhoods and popular housing estates; gated communities and residential complexes; the green of the squares and the reddish hue of the bare streets. In short, encounters and mismatches in a development that contradicts itself by not occurring equally.

2.3.1 - SPATIALISED SOCIETY

The injustice that starts from the citizens towards the city, mentioned by Silva (2003) when he observes exclusion, apartness and socio-spatial segregation, is also present in the space analysed here. According to Silva, the examples of the mistreatment of cities are classic:

> [...] the gradual destruction of equipment, infrastructure and urban furniture, the unbridled migration and cannibalism that chooses new "noble areas" or "expansion areas", to the detriment of others hitherto equipped and suitable for the satisfaction of multiple urban needs. Associated with this unbridled expansion is the accelerated growth of poverty, the emergence of new social subjects and new types of services [...] (SILVA, 2003, p.30-31)

The choice of new areas, as mentioned above, is a proven fact in Jataí when looking at the process of urban expansion. The opening up of these areas, in the face of the great emptiness found in various allotments, expresses an exclusionary action. The "elite" need a space created specifically for them. The part of the population that lacks the resources to live up to ideal standards continues to depend on state assistance in its various forms.

This process of incorporating new plots of land into the urban fabric shows yet another side: the lack of proper planning for land occupation. As seen in a previous study (SILVA, 2002), when analysing land use in a micro-basin within the urban perimeter, areas that do not offer ideal conditions for subdivision have also been added to the city's perimeter. This is the case of the "Córrego do Sapo watershed[12] ". In 1965, 20 per cent of its area was urban. With the process of expansion, the percentage rose to 45% in 1993 and 65% in 2000, generating situations where homes were flooded, such as the one that occurred in 2002, when a house built on the banks of the stream was invaded by water after torrential rain. In the same watershed there are houses situated less than ten metres from the bank of the stream (SCOPEL et al., 2002).

The process of population spatialisation can also be analysed by looking at Figures 25 to 34, which show examples of Jataí's urban occupation.

The city centre, partially seen in Figure 25, the first urban area to be occupied, is today a consolidated area in terms of commerce and services. Due to its date of occupation, which dates back to the beginning of the 20th century, this area was the site of the first imposing residences in the city. The current financial/commercial centre was home for decades to the old elite, descendants of the region's first occupants.

Figure 25 shows two striking points. The first: the Catholic Church headquarters (top left); the second: a residential building (right). Symbols of power and "status" left over from a bygone era, but not completely gone, since it left a consolidated urban space. They are striking symbols of an area of the city that retains its importance more because of its historical occupation process and because of the

[12]This analysis can be found in SILVA (2002).

Figure 25: Jataí-GO: Urban Area - Partial View - Central Sector Ikonos Image, May/2004 - JATAI-GO (2004): Cadastral Maps.

equipment installed in it, than by the urban conditions satisfactory to a contemporary urban centre.

The narrow streets insist on slowing down the pace of the information society. These streets have a greyish tone, which is not characteristic of every urban network. The grey of the asphalt is still a dream for residents of neighbourhoods far from power, but participants in the process of reproducing an unequal urban space.

There are not many empty spaces in this area. The demand for a privileged geographical position that offers good conditions for capitalist activities has transformed a residential sector, by origin, into a locus of services for society. Some houses have simply changed their function, such as the building that houses the city's historical museum (Figure 33).

Walking towards the north of the city, the scenery changes completely (Figure 26). The greyish hue of the streets in the centre gives way to the reddish hue of the bare streets. This area is home to a residential complex donated to the population by the state government and the local town hall. For many, it represents the fulfilment of their dream of home ownership, however simple it may be. For others, it's a "place of mismatch", because its inhabitants don't fit into the capitalist reality that expropriates individuals of their basic rights and values human beings according to their assets.

The neighbourhood in question, called "Colméia Park", which was set up in 1997, brings with it a number of contradictions. The first lies in the term "Park" itself, which is far removed from the standards of quality of life expected in a "residential park", since its namesakes generally offer good urbanisation, with at least a greenish tinge to the trees. However, the green that is found is that of the soya crop (Figure 26 - top right) or even that of the neighbouring forest (located in the micro-basin mentioned above) (Figure 26 - bottom left), which every day becomes more susceptible to suffering the consequences of urban "development", which has already occurred in its southern portion.

Figure 26: Jataí-GO: Urban Area - Partial View - Colméia Park Sector Ikonos Image, May/2004 - JATAI-GO (2004): Cadastral Maps.

One of the riches that sustains the municipality, agricultural production, is at the same time so close and so far away from this area of the city. Close, because it borders the neighbourhood. Far away, because its product doesn't generate benefits for the neighbourhood. The houses built by this wealth are not worthy of a land that has been placed on the margins of society: the poor periphery. They are

are found in more upmarket neighbourhoods, where small, mass-produced houses give way to imposing mansions.

To see them all you have to do is move to the west-central part of the city, a dynamic area that brings back the greyish tone of the streets by adding some blue dots, the pools, which not only reflect the sky, but above all the inequality present in this urban complex (Figure 27).

Figure 27: Jataí-GO: Urban area - partial view of the Planalto Sector. Ikonos image, May/2004 - JATAI- GO (2004): Cadastral maps.

Once again, there are empty lots, unlike the neighbourhood analysed earlier. In this area, the vacant plots are waiting to appreciate in value, which has already been seen elsewhere in the city. Those who can afford the price of unequal development can only enjoy the imposing buildings that express the high quality of the city.

purchasing power of its owners. It is the new elite, crystallising a space considered "noble", far removed from the problems present in much of the city.

They don't, for example, have their homes invaded by the torrential downpours like those that occurred at the end of 2004. The distance between the extremes, between wealth and poverty, is not so great. It's limited to a fence separating the farm from the urban area (Figure 28), or a commercial break on the TV news informing society that production is going well; the rain is the producer's partner. Unfortunately, this is not the case for everyone.

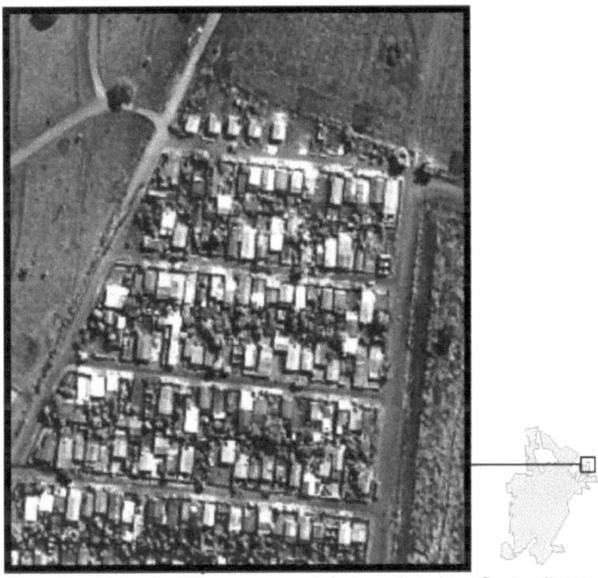

Figure 28: Jataí-GO: Urban area - partial view - Jacutinga Sector. Ikonos image, May/2004 - JATAI-GO (2004): Cadastral Maps.

The darkening of the sky on a rainy afternoon also brings distress to people who don't have a weatherproof home
(when they have a place to live). An individual who is probably part of the figures shown in Table 3: the component of the municipality's HDI with the least expression is income, poorly distributed income.

The city that is sold is not the real one. What's interesting for "urban marketing" is what worked, regardless of the price that was paid, let alone who paid it. Allotments like the one shown in Figure 28, which were originally intended to solve the housing problem, have brought with them other problems.

Set up in the 1990s (in the third period outlined at the beginning of this article) in response to a new need created fundamentally by agricultural expansion in the municipality, these neighbourhoods are basically political in nature, since the number of plots donated was significant, but of poor quality. With an area smaller than that legally required under the land parceling law, these neighbourhoods have densified with a large number of individuals who are unable to afford to build their own homes, either financially or legally. Financial, because the very characteristic of the subdivision is low income. Legal, because the building regulations, established to ensure that the minimum conditions of health and comfort are met, are difficult to follow due to the small size of the plot .[13]

Another subdivision characterised by precarious conditions, especially in terms of urban infrastructure, is located in the western part of the city, in an area earmarked for urban expansion. This is the José Bento Sector. However, expansion took place in a similar way to the Jacutinga Sector (Figure 28). Without offering good living conditions to their inhabitants, these sectors are an example of a lack of planning in urban expansion actions, in a movement that expels the inhabitant to the periphery, since

> Since the price of housing includes benefits in terms of infrastructure and collective means of consumption that make it more expensive than the incomes of the poorest, they are left with the solution of living in distant allotments,

---

[13] The plots in question have an average area of $168m^2$, while the minimum would be $250m^2$ with a minimum frontage of 10m, according to the land parceling law (JATAÍ- GO, 2007).

slums or favelas (PELUSO, 1997, p.237).

In Jataí, these two popular allotments are part of the seven allotments in this category set up in the 1990s, reflecting the high demand for housing, as well as the social inequality present, since the percentage of vacant plots in the city is very high, but they cannot be acquired by people on lower incomes.

Following the route of the spatialisation of society, we come across the sectors Cohacol (Figure 29 - right), built in 1994, and Jardim Goiás (Figure 26 - left), built in 1998 (stage 1) and 2000 (stage 2), comprising the third and fourth periods outlined above. The sectors, located in the northern part of the city, are typically middle-class housing estates with sufficient income to take on a housing loan. These are two of the nine financed housing estates set up in the 1990s and 2000s.

Figure 29: Jataí-GO: Urban area - partial view - northern region.

Ikonos image, May/2004 - JATAI-GO (2004): Cadastral maps.

Moving on, we find the green Diacuy Ecological Park in the centre-south of the urban complex (Figure 30),

Figure 30: Jataí-GO: Urban area - partial view - Diacuy Ecological Park.

Imaoem Ikonos. May/2004 - JATAI-GO (2004): Cadastral Manas.

contrasting with the reddish colour of some vacant lots still present in this part of the city. The result of redevelopment actions, the park in question has increased the value of Jardim Rio Claro, where it is located, and all its surroundings. Situated between a residential area and one with more commercial characteristics, it is an area of special interest for the public administration and is therefore in a zone of urban and landscape interest (ZIUP[14]), established by the city's Master Plan (JATAÍ-GO, 2003).

In the south-west region (Figure 31) we again find a section of the population that doesn't live in an urbanistically correct environment like the previous one. They are people who

[14]See item 3.1.3- Urban areas: the official division.

Figure 31: Jataí-GO: Urban area - partial view - Southwest Region Ikonos image, May/2004 - JATAI-GO (2004): Cadastral Maps.

even though they have equal rights to the inhabitants of the more upmarket areas (at least in theory), they don't live in the same conditions.

Few houses, dirt roads and lots of unoccupied plots are characteristic of this area of the city which, even though it was built in the 1960s and 1970s, still retains traces of its original layout in the 21st century. Located in the least dynamic area of the city, it is practically devoid of public facilities. The nearest school, located in a neighbouring district, only offers primary education, requiring inhabitants to travel further to meet needs such as education and health.

The south-western part of the city requires special attention, especially in terms of occupation. It is characterised as a low-density area, which simply reflects the visibly noticeable voids. Its occupation, like that of the other areas of the city, is necessary for better structuring of the urban whole, avoiding unnecessary growth of the urban area and, consequently, an increase in empty areas.

These aspects show the panorama of a city that is completely different from the one found in 1908 (Figure 32). The more than 90 per cent of the municipality's population who live in the urban area (Table 1), whether through encounters or disagreements, have transformed the soil that originally produced basic agricultural products into an important base for the economy of south-west Goiás. Tables 4, 5 and 6 illustrate the dynamism present in Jataí. It is in the urban area, which is the result of modern agriculture and livestock farming and the dynamism it has created, and which today is gaining autonomy due to the diversification of economic activities (clothing and tourism industries, among others), that inequalities are realised, the clash of opposites, contrasts that may seem simple when you look at them superficially, but which reflect the social structure in force when you read between the lines of events.

Figure 32: Original centre of Jataí-GO - 1908Source: Historical Museum of Jataí (2003) apud MELO, N. A. (2003)

The inhabitants of the early 20th century couldn't have imagined people living under the "protection" of electric fences and high walls. The large windows of the colonial houses, like the one in Figure 33, showed a calmer and safer environment for the population. The houses talked to the city through open doors and windows. Today, the insecurity of the 21st century, the fruit of a "development" that reveals several "side effects", materialises on urban soil the search for security found (at least in theory) in gated communities.

Figure 33 Jataí Historical Museum - Old mansion in the city centre.

Source: JATAI-GO (2004a) - Jataí Dossier - Photo: Gênio Eurípedes

Figure 34 shows the Barcelona Residential Condominium, built in 1998, which reminds us of the reality experienced in today's urban centres. Not just in the big centres, because insecurity has no borders or social classes. This other

The extreme of social inequality, that of the elite, is crystallised in an "object" that is disconnected from the urban whole. A mismatch that reaches the level of academic research, in a space that is closed off, in contradiction to the city that is open to research, as we can see in Silva's (2009) attempt to study this type of city space.

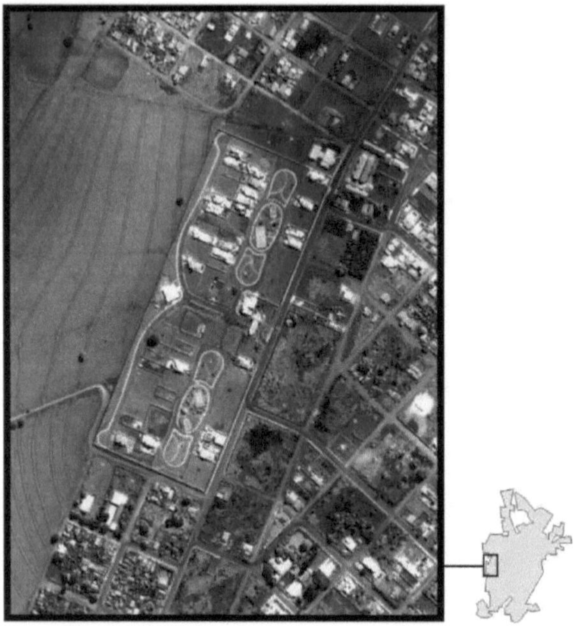

Figure 34: Jataí-GO: Urban area - partial view - Region West - Barcelona Residential Condominium. Ikonos image, May/2004 - JATAI-GO (2004): Cadastral maps.

> [...] when the research attempted to "invade" the private space of the Condominium [...], it was unsuccessful. Although those responsible initially made themselves available and agreed to provide data for mapping the residents, it was never actually made available. After months of contact with the liquidator and the administration in order to obtain the information, they declared that they were afraid of the invasion that a
>
> research could represent in the routine of the condominium owners. According to them, the condominium is a space with its own rules and laws. In their words, "a city within a city, the difference being that we have a higher per capita income than those outside". As such, they said that "residents may not open up their lives to interviews for a survey".

This statement (re)sees this space as one of apartness, or rather self-apartment. In it, the residents want to get away from the life that goes on outside its walls. But in reality, this isn't possible, as the condominium is an exclusively residential space. Residents need to cross its borders to go to other parts of the city, such as schools, shops, churches, clubs and so on. It's not possible for there to be a total lack of contact with those whose per capita income is lower, not least because they enter the condominium as workers.

> They work as gardeners, cleaners, porters, etc (SILVA, 2009, p. 15-16).

In its surroundings you can see several unpaved streets - a common feature of excluded areas - and unoccupied areas - reflecting the process of property speculation. Amongst these landscapes is an object produced by the elite and for the elite which, enclosed within high walls, enjoys a quality of life desired by so many, but within the reach of few.

These different faces of the urban fabric analysed are pieces that, when put together, do not form a homogeneous picture, since the city is not like that. They reflect its characteristics. Its inequalities. The best and worst of this contradictory space.

But with so many contradictions, inequalities and injustices, has the urban fabric really evolved?

In raising this question, it is important to clarify what the real evolution should be.

The reference is to a progressive shift, a real development, in its authentic sense, not to be confused with a simple expansion of the urban fabric (SOUZA, 2003). Development that brings with it a socio-economic transformation, causing the general well-being of the population, raising the standard of living in general (GRAZIANO NETO, 1985) and involving positive social change, thus improving the quality of life and increasing social justice (SOUZA, 2003a).

The possible statements are based on the considerations of Lojkine (1997), when he talks about the effects of urban agglomeration. According to Lojkine, it can be seen that:

> - on the one hand, the useful effects produced by the means of circulation and consumption concentrated in the city;
> - on the other hand, agglomeration effects that are only the

> indirect product of the juxtaposition of means of production or reproduction and are not linked, like the previous effects, to a particular material object. It can be said that their production is, in a way, collective: it is a social combination - conscious or not, deliberate or not - of individual urban agents (builders of housing, shops, offices, factories, etc.). (LOJKINE, 1997, p. 176). (Emphasis in original).

It is a fact that the city of Jataí experienced significant growth, especially in the period after 1970. The "useful effects" of urban growth improved the local quality of life, transforming the small 19th century town into a benchmark for the south-west of Goiás. However, development has not kept pace with growth, resulting in "agglomeration effects" that are at odds with the wealth of the municipality.

The combination of the city's builders has generated uneven development that does not cater for the largest part of the social sphere. The low-income population has come to depend on state assistance for basic human rights. In order to live, they have to wait for the government to give them a slice of valuable urban land. If this happens, the conditions to build the house are lacking. But in due course, you get the building materials. When the building is erected, it houses a family, or even more, who are unable to support themselves. But that's not a problem either, food is part of the "government package", which treats man as a mere number in the public accounts.

Concrete conditions must be promoted so that the city has the same colour tone. Urban centres will not develop if they are simply a jumble of houses and people.

CHAPTER 3

IDENTIFYING URBAN SEGREGATION IN JATAÍ

"All these cities were fortified with high walls, gates and bars... "
Deuteronomy 3:5a

In less than four hours of driving around in a car, observing the urban space that is the focus of this work, and in order to see beyond what is visible, the separations that exist in the urban fabric of Jataí can be seen without too much difficulty. Understanding that the visual distinction of urban areas is merely a reflection of a denser structure that segregates people and social groups, we propose a reflection on the subject.

By settling on urban land, individuals are carrying out an act, conscious or not, of city building. Whether by choice or by imposition (social, economic or otherwise), the settling of individuals in the city's neighbourhoods shapes and crystallises existing social relations. From the point of view that the functioning of urban society selectively transforms places (VILLAÇA, 2001), each place in the city expresses the society that constitutes it.

The term segregation can be understood, according to the Houaiss Dictionary of the Portuguese Language, as "removal, separation, act of isolating oneself or being isolated from others". Or, in terms closer to geography, as a "process by which people establish or localise themselves within a community of areas already occupied by people of similar characteristics or social activities to their own" (INSTITUTO ANTÔNIO HOUAISS, 2002) (emphasis added). A correct understanding of the term gives direction to the understanding of this socio-spatial practice. Its etymology refers directly to the term separation and indirectly to remoteness and deprivation.

From this point, we can see the implications of urban segregation in the city. Separation occurs in the city. In the city, people isolate themselves or are isolated. Like-minded people group together. These actions also lead to the distancing of those who are different and to deprivation, the latter being more dramatic for those who are deprived of experiencing the city fully.

Of the various types of segregation that exist in cities, the segregation of social classes will be discussed. According to Villaça (2001, p.142), this segregation can be understood as "a process according to which different social classes or strata tend to concentrate more and more in different general regions or sets of neighbourhoods in the metropolis" (emphasis in original). Although this study does not directly analyse the separation of Jataiense society into classes, it is understood that the concepts used are pertinent. In this specific case, the classes are "income classes", which gives a very close profile to social classes in terms of what Santos (2008) observes, focusing on the notion of class in understanding systems of economic inequality, in a context of "social conflict based on asymmetries between what people have and what they do with what they have" (SANTOS, 2008, p.355) (Emphasis in original).

Analysing the urban space of Jataí, we can see that it is similar in some respects to the pattern of segregation described by Villaça (2001) in the case of the Brazilian metropolis: centre and periphery.

> The first, equipped with the majority of public and private urban services, is occupied by the upper classes. The second, under-equipped and remote, is occupied predominantly by the excluded. Space acts as a mechanism of exclusion (VILLAÇA, 2001, p. 143).

In the case of Jataí, there are different locations, equally distant from the city centre, which have residents with different incomes, but predominantly lower incomes. Jataí differs in scale, it is a city classified as a "Zone A Centre" (IBGE, 2007b), a medium-sized city in its region; but it presents the problem of socio-spatial segregation.

Segregation is dynamic, it involves elements of space and time, and has different speeds (CORRÊA, 1995). This dynamic is characteristic of capitalism, and its rhythms and directions are imposed by it. The urban fabric presents space, the periods of occupation of areas of the city reveal time and speeds that occur according to the demands of each period (economic, social and political).

The most dynamic economic activities have favoured locations with the best infrastructure for their installation. Within this same capitalist logic, people with greater purchasing power will live in places that meet their "needs". Places "lose" and "gain" value. Areas, both residential and commercial, are produced to serve the interests of the ruling class, both locally and globally.

The value of the property

> is created by a socio-spatial matrix of locations and activities associated with the production of wealth, which then gives particular urban sites a use value that is more a function of that socially determined spatial pattern than of an intrinsic quality of the resource itself. (GOTTDIENER, 1997, p.179)

This reality is clearly seen in the city centre, where the densification of commerce and services has raised the price of urban land, consolidating the central area as a place restricted to those who pay its high price. In this area, the privileged location as the city's original centre and the production of wealth have been combined in a direct link with the dynamics of this region. Thus, high property values remain: "use values produced by the spatial matrix of capital accumulation activities [translated] into commodity exchange values reflected in property prices" (GOTTDIENER, 1997, p.179).

Why does segregation occur in this way? Why does it occur in certain places and not in others? What are the reasons for choosing prime or depreciated areas? What causes segregation? To live near your equals? To have social status? These questions, raised by Villaça (2001), are pertinent to this discussion. Let's think about them.

3.1 - URBAN INFRASTRUCTURE

Lojkine (1997) raises other aspects of socio-spatial segregation: segregation at the level of housing; segregation at the level of public facilities (schools, nurseries, sports facilities, social facilities...); and segregation at the level of home-work transport. Thus, the income of individuals is the strongest component of access to the city, in a scenario in which "[...] socio-spatial interests divide into fractions not only the capitalist class but also the working class" (GOTTDIENER, 1997, p.167).

The analysis of this urban complex is based on the confrontation of various variables. In this sense, an important aspect of the consolidation of Jatai's urban space is the infrastructure available. According to Lojkine (1997, p.180)

> the city, the region and the various types of spatial agglomeration would be the combination of infrastructures that are partly inseparable and closely complementary, providing an indispensable basis for the different activities.

In this passage, Lojkine, dealing with the capitalist limits of urbanisation, raises some factors that are directly linked to the economic evolution of cities. These include:

> * the training infrastructure (all the means of general education and vocational training, at all levels, on which the quality of the region's human potential depends);
>
> * socio-cultural infrastructures (housing, health, cultural and leisure facilities);
>
> * the economic infrastructure itself, which mainly includes: industrial estates, roads, energy and public transport networks, etc.
>
> telecommunication (LOJKINE, 1997, p.180). (Emphasis in original).

These three items would be integrated into the different levels of the regional space:

> * the company's infrastructure (industrial estates), including the purchase of land, common equipment and possibly pre-construction;
>
> * urban infrastructures comprising "structuring" facilities (public centres), "accompanying" facilities (residential areas and links to networks) and intermediate facilities (social centres);
>
> * local infrastructure for industrial use: school facilities, toilets, local transport equipment;
>
> * regional infrastructures: universities, research institutes and organisations, regional transport and telecommunications networks (LOJKINE, 1997, p. 180-181). (Emphasis in original).

These considerations bring us back to Jatai's urban space in order to see the role of some of these aspects in our study area. In this way, it is possible to have a broader view of the existing urban ensemble. It is therefore important to consider some of these factors in this analysis.

3.1.1 - TRAINING INFRASTRUCTURE

In terms of educational infrastructure (Figure 35), offers, as already described in section 2.1 (Jataí's insertion into the Brazilian urban network), public and private educational establishments ranging from primary to higher education.

With regard to primary and secondary education, public schools are more widely distributed in the urban area. The majority of neighbourhoods are served by schools, but these are concentrated in the central-western area of the city, with the south-western region of the city having the fewest schools, a fact directly linked to the lower dynamism found in this region (Figure 22). The private units of this type of education are primarily located in the central region, with a tendency to move towards the west of the city.

Higher education establishments, both private and public, are concentrated in the west of the city, catering for students from both Jataí and other neighbouring municipalities. The exception is the UFG Campus, located in the far south-east of the city.

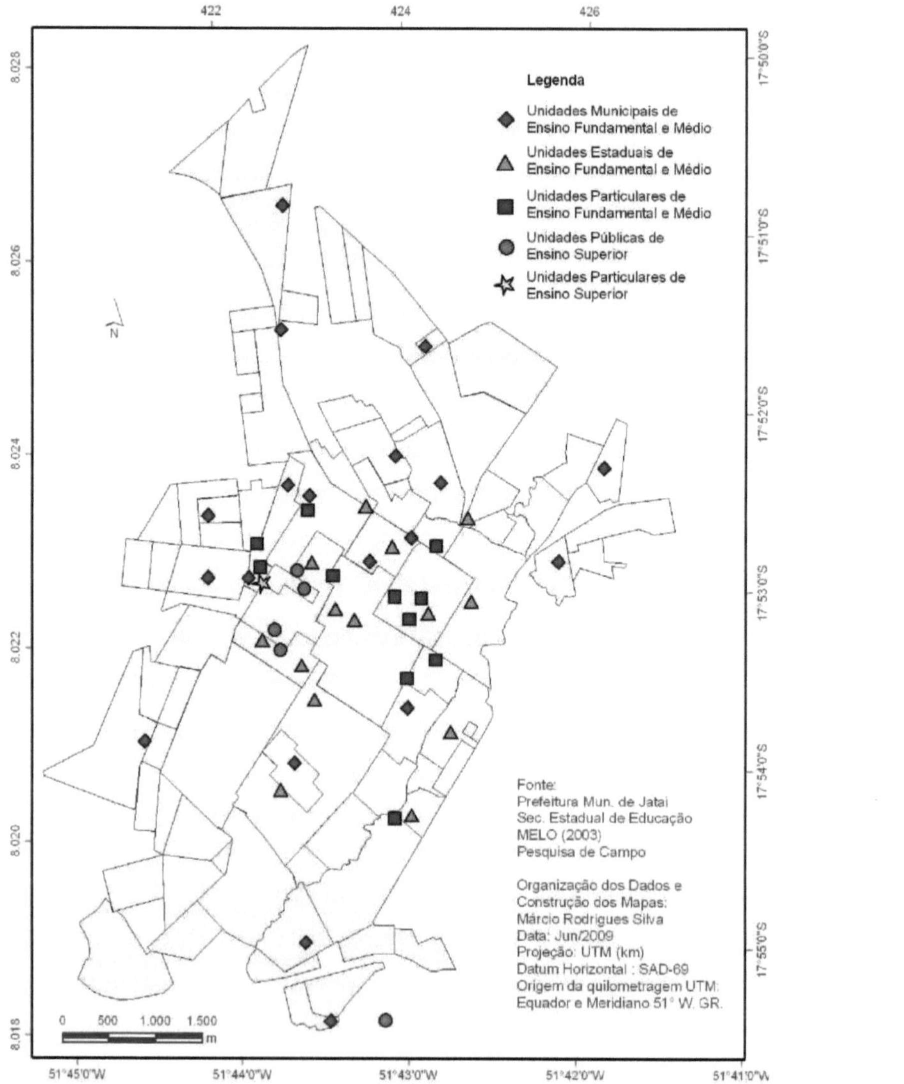

Figure 35: Jataí-GO: Location of educational establishments

3.1.2 - PUBLIC SERVICES, HEALTH AND LEISURE

The movement analysed in the city's dynamic areas expresses the current trend of local urban expansion. It is then possible to complement this analysis by observing the distribution of some socio-cultural facilities.

In this context, the location of the main public utility establishments deserves special attention (Figure 36). Similarly to economic activities, these facilities appear in the city's original core and move westwards as a priority. In this way, areas that are already more dynamic, especially due to the presence of shops and services, are strengthened by the movement generated by public utilities that periodically attract the population that needs them.

The socio-cultural needs of the population are therefore one of the driving forces behind the growth of the city. If there is a production of the city, and of social relations in the city, it is a production and reproduction of human beings, rather than a production of objects (LEFEBVRE, 1991, p.47).

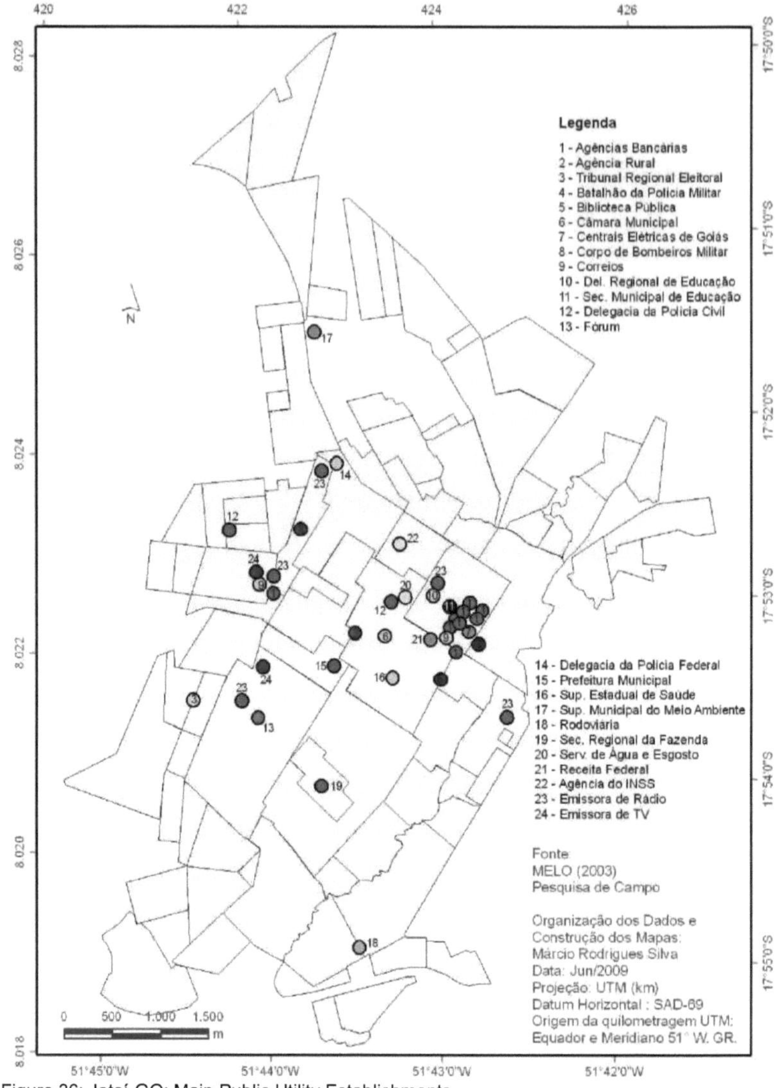

Figure 36: Jataí-GO: Main Public Utility Establishments

Since Jatai society is part of the process of capitalist reproduction, the areas that receive urban expansion are born within the logic of this process. New allotments are set up because there is a demand for housing. The new inhabitants need to eat, have fun, study, need health care, etc. To cater for them, new facilities are installed: a local market, a larger one that attracts other related activities. In short, in this incessant movement, the region that best responds to the demands placed on it is consolidated.

The government installs a school, a health centre; benefits are implemented. Regardless of how they are achieved, or how long they take to be implemented, what is certain is that they add value to a region and even generate "pride" on the part of local residents who communicate in an urban neighbourhood that is not identified with the metropolis or a village with typically rural features.

In this sense, the urbanite reveals that

> The city has always had a relationship with society as a whole, with its composition and functioning, with its constituent elements [...] with its history. It therefore changes when society as a whole changes. [...] the city also depends, and no less essentially, on relations of immediacy, on direct relations between the people and groups that make up society [...] (LEFEBVRE, 1991, p.46).

The urban environment demands space, imposes rules, generates encounters and disagreements, joys and sorrows, dreams and nightmares. Whether they meet or not, people want to have fun or simply forget about their day-to-day problems. Figure 37 shows the main cultural and leisure options on offer in the city. With the exception of the Tourist Pole, located approximately 10km from the city and requiring a longer commute, the majority of the population can count on public squares, which are few, ecological parks and clubs and other private leisure facilities. Access to the latter two depends on the purchasing power of the population, making the public parks (urbanised artificial lakes) places of great visitation.

In the health area (Figure 38), there are basic health units distributed in such a way as to serve the local population without too much travelling. However, hospitals are located in the central and western areas of the city, requiring more travelling, depending on the user's needs. Private healthcare establishments are concentrated in the city centre, occupying the first dynamic regions delimited above. The trend observed is that, like other activities, they have to move in a westerly direction, primarily in the vicinity of the municipal health centre located in this area.

Of the hospitals, only one is public. It concentrates all the care outside the reach of the local units (Family Health Units), not only in the municipality of Jataí, but also in the neighbouring towns it polarises.

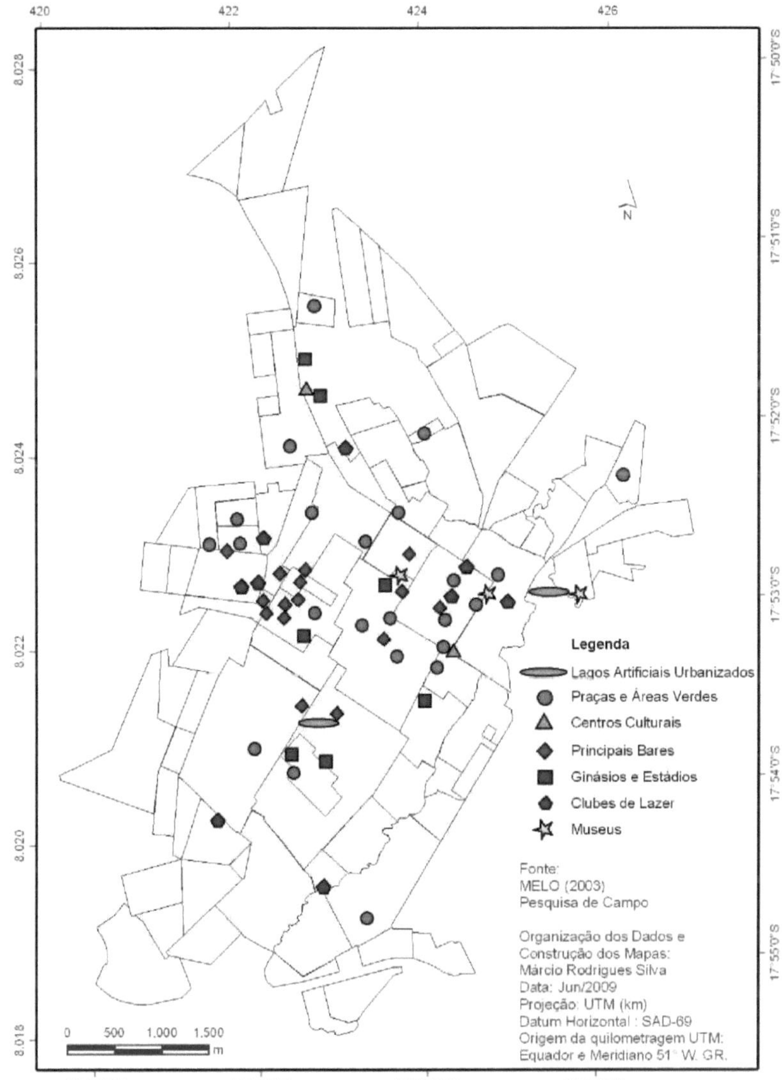

Figure 37: Jataí-GO: Main Culture and Leisure Options

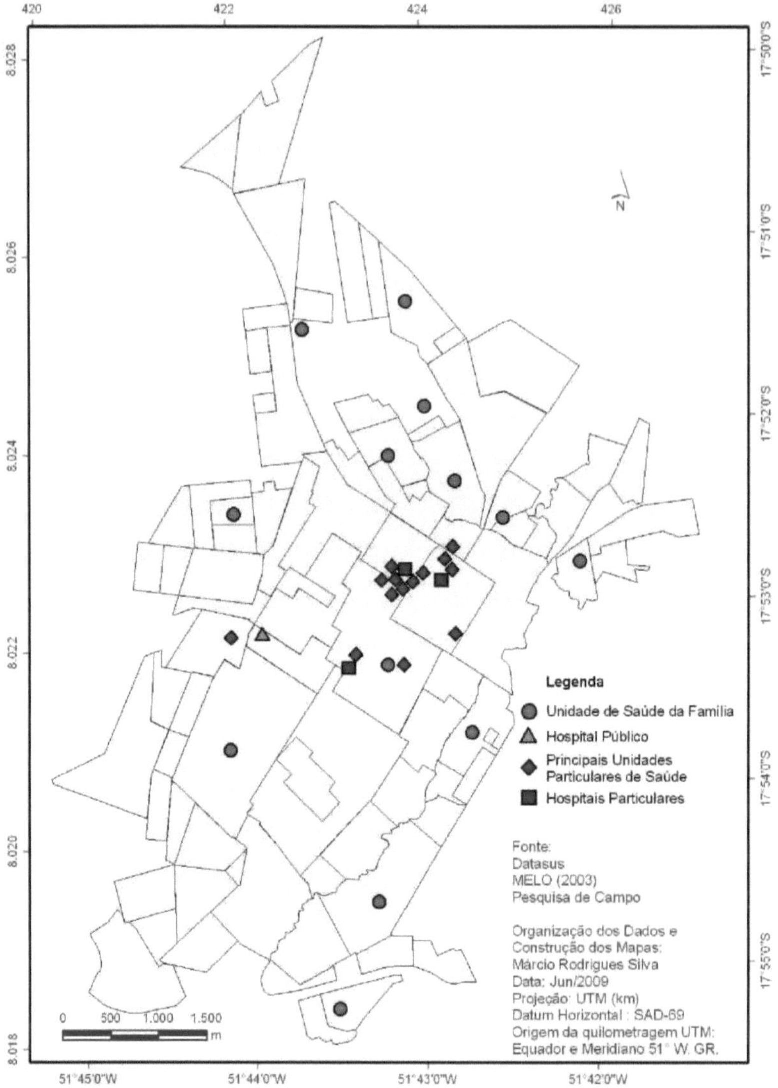

Figure 38: Jataí-GO: Main Health Establishments

3.1.3- URBAN AREAS: THE OFFICIAL DIVISION

Urban land use in the city of Jataí had its parameters established with the implementation of the Municipal Master Plan in 2001. This was revised in 2003, 2005, 2007 and 2008. Currently, the urban zones (JATAÍ, 2008) are delimited according to municipal law no. 2861/08, of 18 April 2008, as shown in Figure 39. They are:

§ 1 to § 8 - Residential Zone I - ZR I

- An area where single-family homes are predominantly concentrated and where other uses are considered complementary and accessory to this urban space. It is characterised as a low population density zone of up to 100 inhabitants/ha.

§ 9 to § 14 - Residential Zone II - ZR II

- This is an area where single-family or collective residential homes and buildings are predominantly concentrated, and where other uses are considered complementary and accessory to this urban space. It is characterised as a zone with a medium population density of 100 to 500 inhabitants/ha.

§ 15 - Residential Zone III - ZR III

- This area is predominantly home to single-family homes and residential buildings for low-income populations or housing programmes with plots of less than 360.00 m^2 (three hundred and sixty square metres), and where other uses are considered complementary and accessory to this urban space. It is characterised as a zone with a medium population density of 100 to 500 inhabitants/ha.

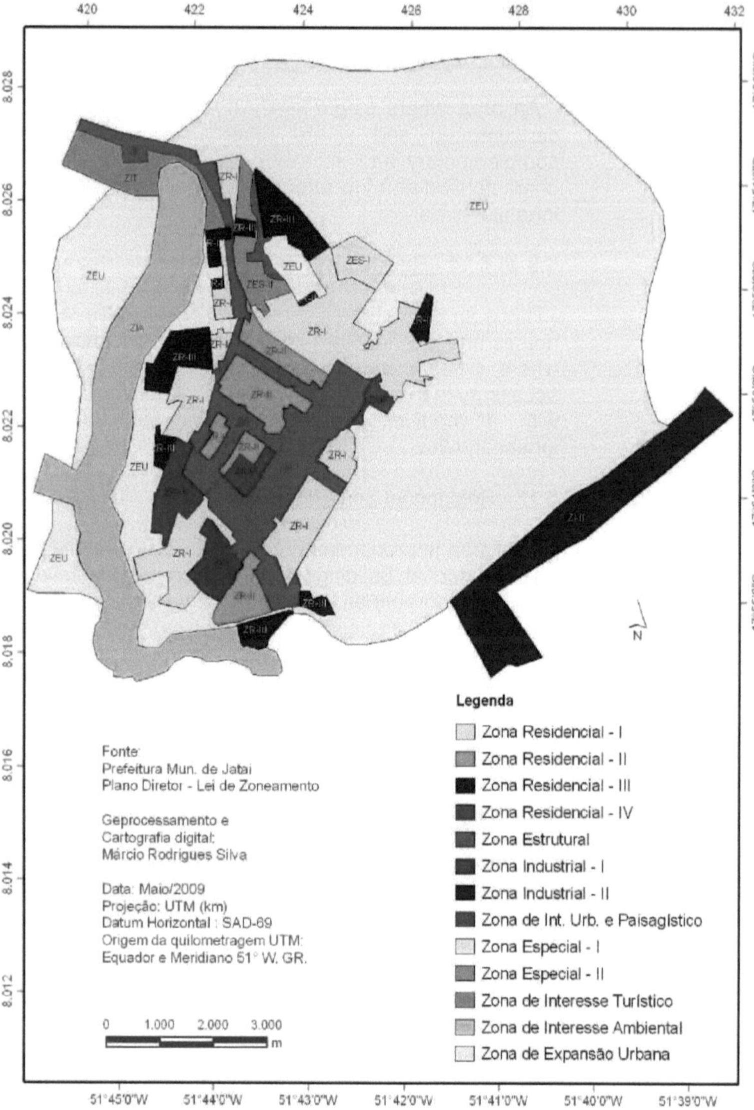

Figure 39: Jataí: Urban Zoning - 2008

§ 16 - Residential Zone IV - ZR IV

This is an area where single-family or collective residential homes and buildings are predominantly concentrated, and where other uses are considered complementary and accessory to this urban space. It is characterised as a medium/low population density zone of 100 to 400 inhabitants/ha.

§ 17 to § 19 - Structural Zone -ZE

- This is an area where commercial and service activities are predominantly concentrated, and where other uses are considered complementary and accessory to this urban space. It is characterised as a zone with a medium population density of 100 to 500 inhabitants/ha.

§ 20 - Industrial Zone I - ZI I

- Areas where industries with a low degree of environmental degradation are concentrated and where other uses are considered complementary and accessory to this urban space.

§ 21 - Industrial Zone II - ZI II

- Areas where industries with a medium to high degree of environmental degradation are concentrated and where other uses are considered complementary and accessory to this urban space.

§ 22 - Environmental Interest Zone - ZIA

- Protection areas, which due to their characteristics should be preserved and/or revitalised through appropriate measures in each case, with the occupation of their surroundings strictly controlled by the public authorities.

§ 23 - Special Zone I - ZES I

- These are areas that, due to their particular neighbourhood and occupation characteristics, must have their occupation controlled in order to guarantee their viability without damaging the environment. They are characterised by areas intended for the implantation of farms or recreational sites that have already been implanted, and dismemberment for individual housing or industrial implantation is not permitted.

§ 24 - Special Zone II - ZES II

- These are areas which, due to their particular neighbourhood and occupation characteristics, must have their occupation controlled in order to guarantee their viability without damaging the environment. They are characterised by areas intended for the establishment of recreational and/or social clubs, already established industries and existing public facilities or buildings; dismemberment for individual housing or new industrial establishments is not permitted.

§ 25 e § 26 - Zone of Urban and Landscape Interest - ZIUP

- These are areas that, due to their particular neighbourhood, occupation characteristics and the presence of certain public facilities, should have their occupation controlled in order to guarantee the viability of their implementation without harming the quality of the surroundings, urban infrastructure and the urban environment, as well as the execution of consortium urban operations.

§ 27 to § 29 - Urban Expansion Zone

Empty urban area preferably destined for the subdivision of urban land or the implementation of public facilities and consortium urban operations.

§ 30 and § 31 - Zone of Tourist Interest - ZIT -1

- These are areas that, due to their particularities, are subject to the implementation of public or private facilities, urbanisation of leisure areas, as well as consortium urban operations aimed at the implementation of tourism and leisure elements. (JATAÍ, 2008)

According to these regulations, land occupation is subject to legal specifications. For the most part, urban areas are made up of mixed residential areas, involving neighbourhood commercial establishments to meet local demand.

In the context of this analysis, we highlight the areas included in ZR III, which, due to their population density characteristics, have their own economic/social dynamics, thus requiring specific policies to cater for these localities. ZR-III generates a large flow of people to areas that meet their demands, be it for work/employment, shopping, leisure and/or public services. This movement is mainly towards the more centralised areas.

The division into zones demonstrates not only the boundaries stipulated for urban land and its use. Society is also officially divided. Individuals with lower purchasing power are moved to the outskirts of the city, where the highest population density is already concentrated, and which tends to increase with the legal determinations. This division goes beyond the limits of urban space and reveals the existing social division.

It does, because the division into zones reflects the city that existed before the zoning law. Most of the zones were defined according to the reality found at the time the law was drawn up. As well as revealing, it consolidates the division; the zoning law specifies that these areas are destined for the low-income population (JATAÍ,

2008). This is the place of the poor.

In the general provisions of this law there is also a contradiction with the guidelines of the masterplan, which leaves loopholes for bad faith actions. What it currently reads is:

> Art. 1 - The use and occupation of urban land in the entire urban area of the city of Jataí will be governed by this law.
>
> Art. 2 - The objectives of this Law are:
>
> 1 - Establishing criteria for the occupation and use of urban land;
>
> 2 - Control population densities in order to establish orderly growth;
>
> 3 - To make different uses and activities compatible within certain parts of the urban space in order to observe the proximity of incompatible or inconvenient uses;
>
> 4-0 land subdivision, building or use that is excessive or inappropriate in relation to the urban road and infrastructure;
>
> 5 - The installation of equipment or activities that could act as traffic-generating centres, without the provision of corresponding infrastructure;
>
> 6 - Speculative retention of urban property that results in its underutilisation or non-utilisation;
>
> 7 -The deterioration of urbanised areas;
>
> 8 - Pollution or environmental degradation (JATAÍ, 2008). (emphasis added).

The interpretation is that the objectives of the law are also to generate problematic situations for the city. This law has the following objectives: ...excessive or inadequate use in relation to the urban road and infrastructure; the installation of equipment or activities that could act as hubs generating traffic, without the provision of a corresponding infrastructure; the speculative retention of urban property that results in its under-utilisation or non-use; the deterioration of urbanised areas; pollution or environmental degradation.

The guidelines of the masterplan include:

9 3 - *The* objective of the Jataí Urban Master Plan is to guarantee a balance between demographic/economic growth and the quality of life and well-being of the population in the urban area defined in its perimeter limits by specific law and divided into zones of characteristic uses, having as reference, in addition to the Federal Constitution, Law n. 10.257 - of 10 July 2001, which establishes the City Statute.

10 4 - The social function of the city is understood as ensuring the general conditions for the development of production, commerce and services, and particularly for the full realisation of citizens' rights, such as the right to health, education, basic sanitation, work and housing, public transport, safety, information, leisure, environmental quality and participation in planning.

11 5 - The social function of urban property is understood to be the function corresponding to the activities carried out on each property when the use and occupation of each property, whether of public or private domain, responds to the requirements that fulfil the social function of the city, consolidated in the guidelines of the Urban Master Plan and in accordance with the legal provisions deriving from it.

12 6 - Urban property fulfils its social function when it meets the fundamental requirements of urban planning, ensuring that citizens' needs are met in terms of quality of life, social justice and the development of economic activities, through the following general guidelines:

1 - guaranteeing the right to urban land, housing, environmental sanitation, urban infrastructure, transport and public services, work and leisure;

2 - democratic management through the participation of the population and associations representing the various agents of the community and organised segments of the community in the formulation, execution and monitoring of development plans, programmes and projects urban;

3 - Cooperation between the levels of government, private society and other sectors of organised society in the urbanisation process, in the social interest;

4 - Planning the development of cities and the spatial distribution of the population in order to avoid and correct the distortions of urban growth and its negative effects on

the environment;

5 - Ensure an adequate supply of urban and community facilities, transport and public services suited to the interests of the community and local characteristics;

6 - Planning and control of urban land in order to avoid:

6.1 - The inappropriate use of urban property;

6.2 - The proximity of incompatible or inconvenient uses;

6.3 - Land subdivision, building or excessive or inappropriate use in relation to the urban infrastructure;

6.4 - The installation of developments or activities that could act as traffic-generating centres without the necessary infrastructure;

6.5 - The speculative retention of urban property, resulting in its under-utilisation or non-utilisation;

6.6 - The deterioration, pollution and degradation of urbanised areas or the natural environment.

7 - Adaptation of economic, tax and financial policy instruments, as well as public spending, in order to guarantee the objectives of urban development, favouring investments that generate general well-being and its enjoyment by the different segments of society.

8 - Land regularisation and urbanisation of areas occupied by low-income populations through the establishment of specific urbanisation standards, taking into account the socio-economic situation of the population and environmental standards.

9 - Fair distribution of the benefits and burdens arising from the urbanisation process. (LAW No. 2.804, OF 22 JUNE 2007). (JATAÍ, 2007) (emphasis added).

Taking care of the legal wording is basic, in order to support the actions of public authorities. It is hoped that the social function of the city is fulfilled, that the law is correct and that it is correctly applied. But what we see is not yet the case.

Once again, occupation of the city's vacant areas becomes

distant. The more central areas, which concentrate a greater number of public facilities, produce an increase in the value of the properties located there. This is the case with properties located around the ZIUPs, areas where ecological parks have been built

The areas with the highest vacancy rates (Figures 11 and 12) are mostly located in the ZR-I, ZR-II and ZR-IV zones, which are characterised by single-family residential use and by low

demographic density. These areas reveal the still high cost of urban land, since they have been left aside in terms of urban occupation, becoming intermediate areas between more densely populated central and peripheral locations. They thus contradict what the Master Plan stipulates: the ordering and control of urban land in order to avoid: [...] the speculative retention of urban property, resulting in its under-utilisation or non-utilisation (JATAÍ, 2007).

The new economic activities tend to follow the delimitation of the Structural Zone due to legal determinations, as well as the infrastructure present in it, especially travel routes (main streets and avenues).

The priority urban expansion areas follow the trend described in the masterplan:

> To direct the city's growth, in the medium and long term, towards the western region of the urban perimeter, which offers good topographical conditions for urbanisation and with the corresponding expansion of the road network and basic infrastructure.(LAW No. 2.804, OF 22 JUNE 2007, Chapter II, Article 3, Subparagraph 8). (JATAÍ, 2007). (Emphasis added).

Safeguarding the preservation areas which, together with the military area present in the city (most of the eastern urban expansion zone), form a perimeter limiting new urban occupations, the city grows westwards. Thus, the larger industrial activities located in the LZ are separated from the already consolidated urban area; a factor favouring occupation of the LZ is the access roads present in its surroundings: BR's 060 and 364.

3.2 - SEGREGATED SPACES IDENTIFIED

Although high-end housing can be found in neighbourhoods that offer cheaper plots, the process of segregation and self-segregation in Jataí is directly linked to the price of urban land. Within a perverse capitalist logic, segregation influences the price of land, since it increases the value of intermediate areas, and reproduces the process, again expelling lower incomes to cheaper areas.

Looking at Figures 42 and 46, the concentration of the upper class in the most expensive areas of the city is clear. On the outskirts, lower-income individuals predominate. On the map, the points that contrast with the rule of the peripheral neighbourhoods show the presence of "strangers" to the neighbourhood, in a movement in which the aggregation of some causes the segregation of others Villaça (2001).

In order to gain this insight, it was crucial to spatialise some of Jataí's social information. The procedures needed to create this database are presented below.

3.2.1 - MAPPING SEGREGATED SPACES

The data used refers to the IBGE's Information Base by Census Sectors for the year 2000, available in "Shapefile" format. These were imported into ArcGIS® software and combined with the urban map of Jataí (Figure 40), in order to obtain results according to the actual layout of the neighbourhoods, since the boundaries of the census sectors do not coincide with the municipal urban network.

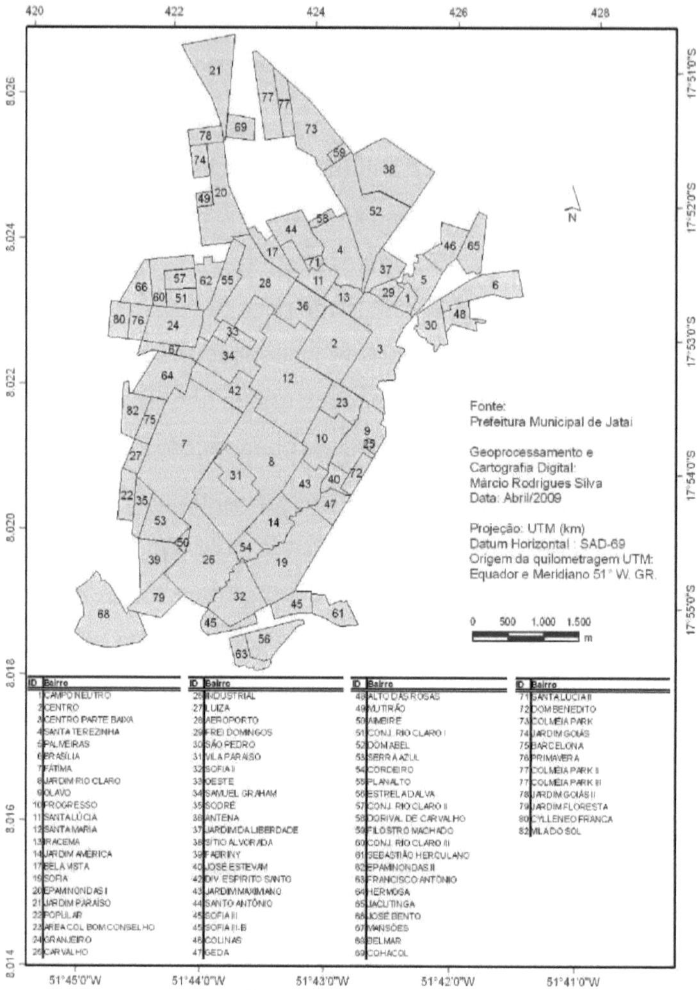

Figure 40: Jataí: Urban area - 2004

To create the "Urban Segregation" map (Figure 42), used the "symbology>quantities>dot density" tool (Figure 41). The information fed into the system refers to the values of *"People responsible for permanent private households - nominal monthly income"*, with income ranging from "0" (people with no income) to "more than 20 MW" (more than twenty minimum wages).

Each point on the map corresponded to an individual, on a

scale of ordered colours, allowing analysis of the quantity and concentration of the population in each neighbourhood, and in the urban area as a whole, according to their income.

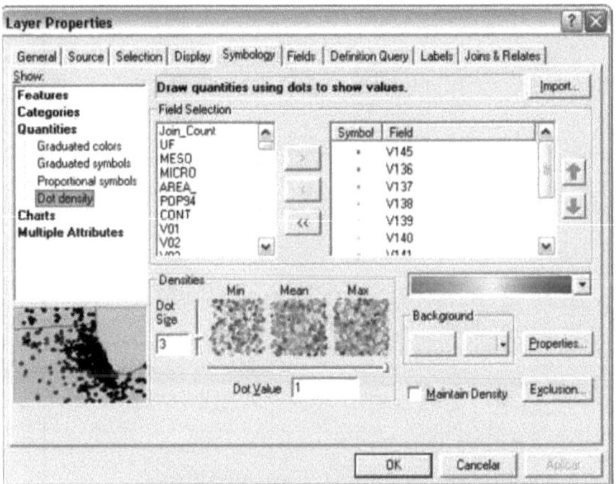

Figure 41: Making the Urban Segregation Map

The proposed study necessarily involves understanding how the population is distributed in Jataí's urban layout. Analysing the "Urban Segregation" map (Figure 42) allows us to add more necessary information to the discussion of this space.

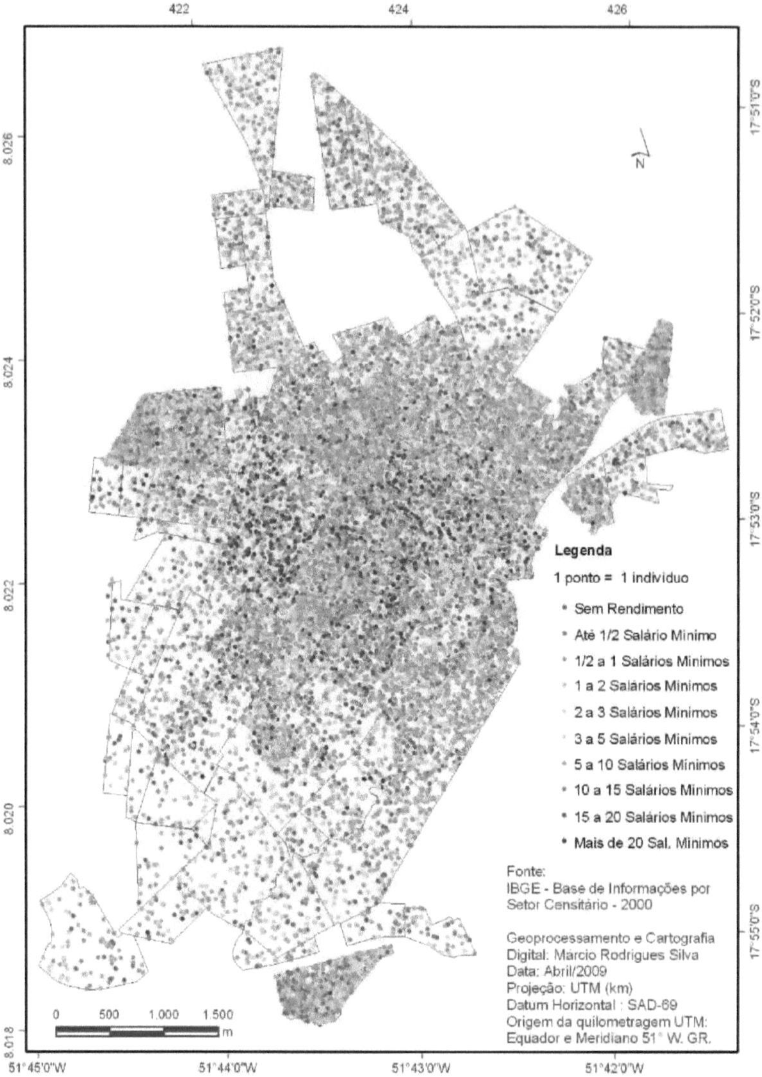

Figure 42: Urban Segregation in Jataí - 2000

Following the hierarchy defined in the legend, it can be seen at first glance that the shades of red indicate the highest incomes and the shades of green the lowest. It's possible to see a concentration of incomes in the more central areas and the lowest values in the more distant neighbourhoods.

Some areas stand out on this map (Figure 43):

- <u>Area 1: Original city centre</u>

The concentration of the higher-income population in the city centre is clear. This place, where the first families from Jatai originally settled, is still home to a large proportion of the highest incomes. In addition to the historical ties and identity with the place, the centre is very easy to access and has a high cost of urban land, which would explain the predominance of high values.

- <u>Area 2: central-western part of the city - upmarket neighbourhoods</u>

In this area, the gestation of a new centre is considered to be taking place, and

> Although the central region appears to be indispensable to the city's dynamism, it is clear that there are movements that tend to move the decision-making centre to areas that offer better access conditions (mainly roads) and greater prospects for development [...] (SILVA, 2005, p.71).

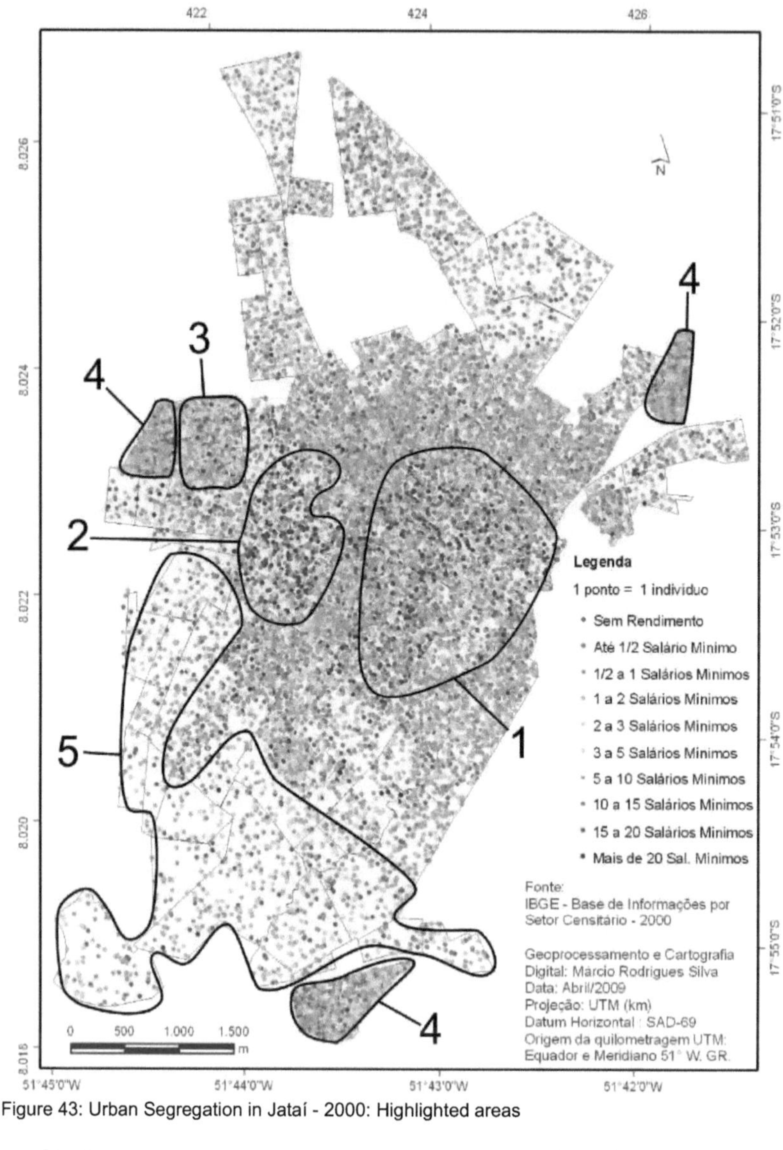

Figure 43: Urban Segregation in Jataí - 2000: Highlighted areas

Similarly, urban land in this area is inaccessible to lower incomes, generating the concentration of an elite that acquires plots not only for housing, but also for property speculation, which can be seen in the 25 to 50 per cent of empty plots (SILVA, 2005).

- Area 3: Middle-income housing estate

This area corresponds to a funded housing estate, basically concentrating a middle-income population, according to the data used.

- Area 4: Low-income housing estates

The three areas highlighted correspond to low-income housing estates, two of which were donated to the residents. Reflecting their initial characteristics, they accommodate the lower-income population, crowded into small spaces, as can be seen in Figure 43.

- Area 5: Area with large urban voids - over 50%

The highlighted area draws attention not because of the concentration of low-income families, but because of the large urban void.

Neighbourhoods in this region have at least 50% of their plots unoccupied, reaching over 90% in some areas (Figures 11 and 12).

In the current decade, there has been a movement towards the peripheral areas of the city with new housing developments (Figure 9). Since the intermediate areas have been allotted, which doesn't necessarily mean they are occupied, the target of the allotments is the areas furthest from the central core.

In 2007 alone, three large allotments were opened in Jataí. This greatly increases the number of vacant plots and contributes to an even greater increase in the value of the intermediate areas, which are already so distant from the low-income population.

However, there are no effective public policies to integrate the segregated population in terms of access to the goods and services necessary for a good quality of life, and in accordance with the urban masterplan.

It is necessary to insist: actions to make better use of urban space would materialise development in Jataí in its broadest sense, "[...] development understood as positive social change [...]" (SOUZA, 2003a, p.60). As long as the population's priority needs are not met, the city's social function will not be fulfilled.

3.2.2 - GOVERNMENT ASSISTANCE AND SEGREGATION

The information on government assistance cooperates in this analysis, highlighting yet another element that reveals the segregation present in the city. Analysing socio-spatial segregation specifically through data from the Bolsa Família programme, we can see a similar pattern of income distribution in the city. The data used refers to the period from 2005 to 2008.

Since 2005, various assistance programmes have been concentrated in the Bolsa Família. In this way, these data include parts of the population that, as well as not having incomes declared in the IBGE survey, depend on government transfers.

Figure 44 shows this data through concentrations similar to the IBGE income map, which confirms the presence of individuals with lower incomes on the outskirts of the city, especially in the areas already mentioned.The areas numbered in Figure 45 are similar to those highlighted in the previous item. The neighbourhoods located in areas 1, 3, 4 and 6 are low-income housing developments. All the areas coincide with low income levels, as shown in Figure 44. Area 2 is a group of older neighbourhoods, built up until the 1980s, which also have low incomes, despite not having been conceived as "low-income housing estates".

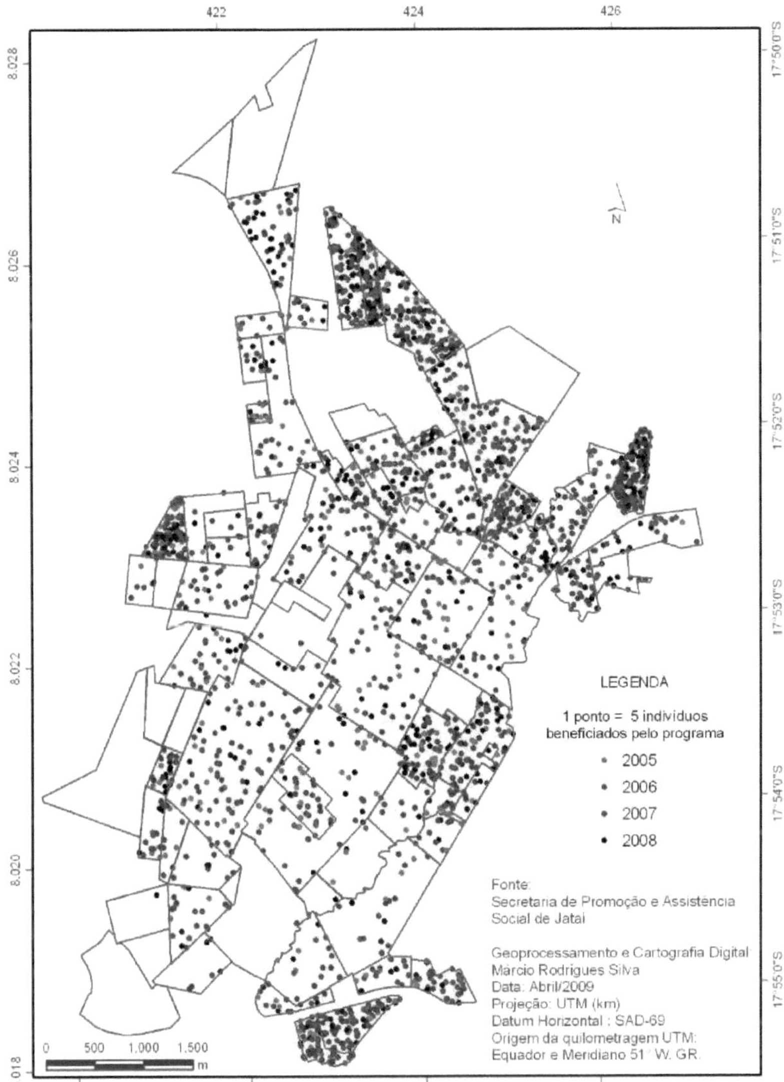

Figure 44: People Benefiting from the Bolsa Família Programme - 2005 - 2008

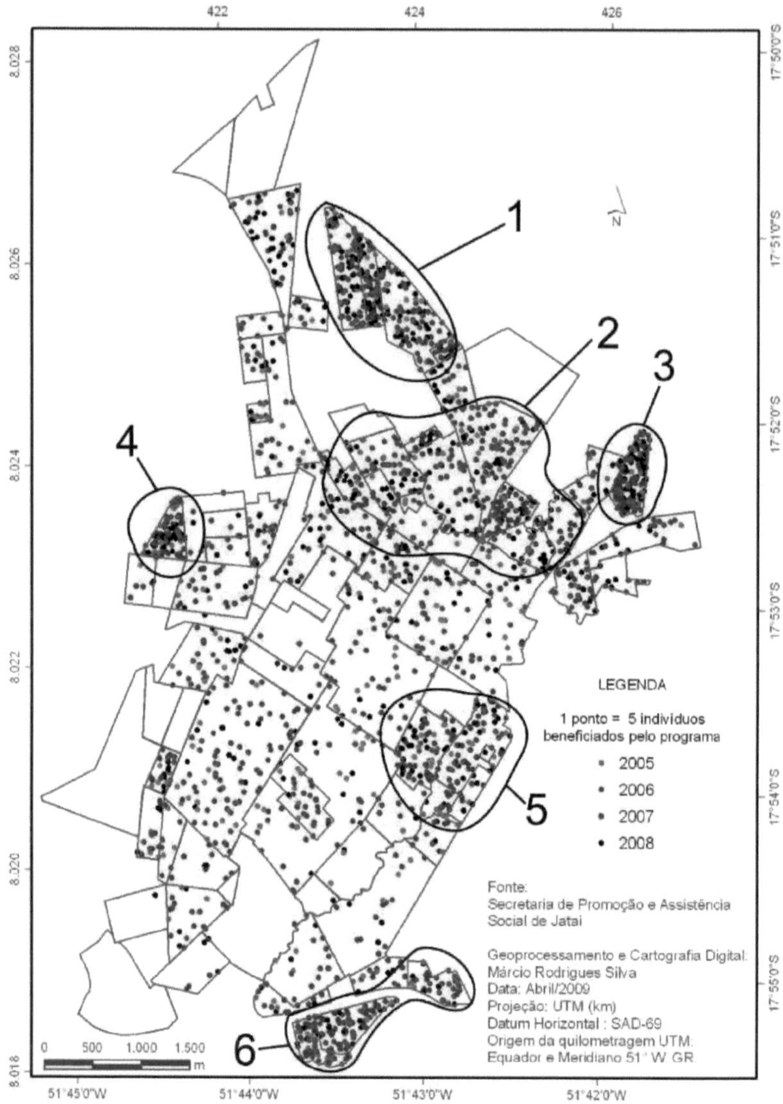

Figure 45: People Benefiting from the Bolsa Família Programme - 2005 - 2008 - Highlighted Areas

In this sense, it is clear that the population with lower incomes lives predominantly on the outskirts of the city. This can be confirmed by analysing the construction standards in these areas.

The lower the income, the worse the standard of the buildings, which reflects a distance from a good quality of life. Figure 46 shows the building standard in Jataí, and once again the periphery is synonymous with exclusion. In contrast to the "medium" and "high" standard buildings, the "precarious" standard surrounds the urban fabric.

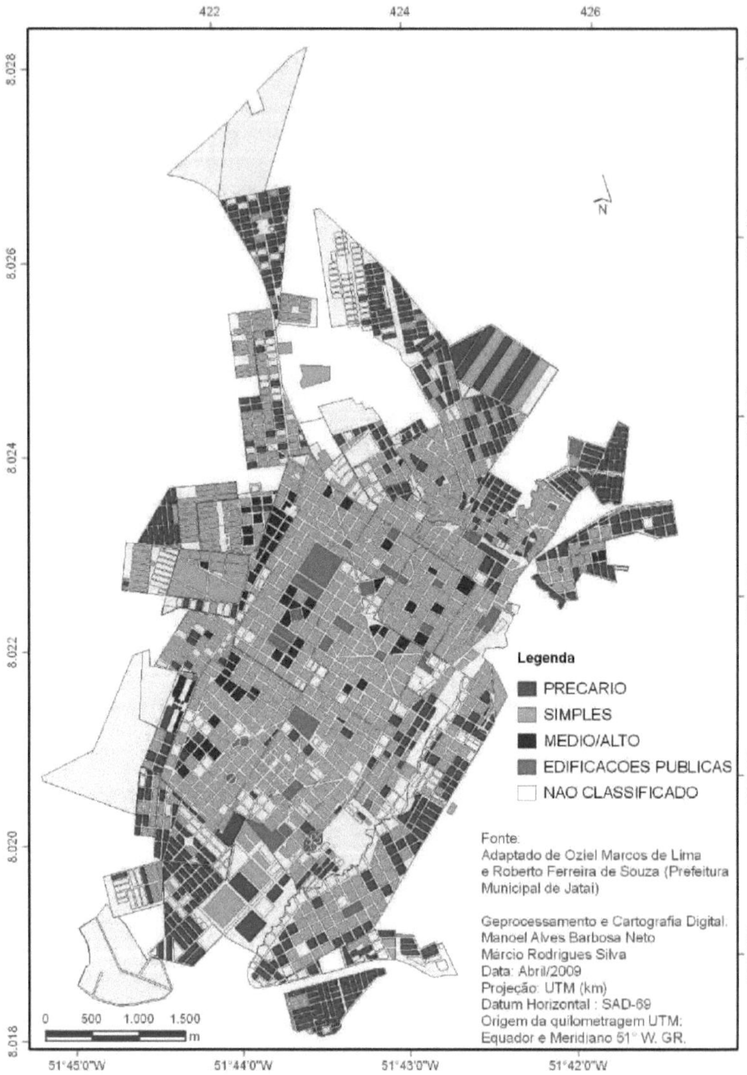

Figure 46: Jataí - Building Standard - 2007

This space, most of which is residential, is on the one hand historically constituted and on the other articulated with the social structure as a whole (CASTELLS, 1983). The historical constitution refers to the context of modern agriculture, especially since the 1980s. The articulation with the social structure makes visible in the city a separation between "equals", which reflects the oppressive context of most of Jataí's peripheral areas. Its inhabitants are not part of the official city. They have their own city. In Kaztman's (2005) view, they are seduced by a modern society; they can only participate symbolically and cannot overcome the obstacles to achieving equivalent material participation in the city that is publicised on TV commercial breaks.

3.3 - THE PERIPHERAL GAZE: A DIFFERENT VIEW OF THE CITY

Analysing the landscape is fundamental in the process of understanding urban space. Understanding landscape as one of the ways in which geographical space manifests itself, which goes beyond the idea of landscape as a static element, means seeing the city through its formal manifestation (CARLOS, 1994).

Understanding the production of the visible aspect of space, which expresses the content and the social relations that made it up, within a historical, social and concrete context[15] , requires different perspectives. This analysis must, in the words of Sauer (1998, p.17), discover the connection and order of the phenomena that make up a given area. To do so, we must consider that these phenomena are associated or interdependent, because "[...] space is a mixture, a hybrid, a compound of forms and contents" (Santos, 1999, p.35).

The city, complex by nature, demands bold investigations that aim to reach it. Such is the city:

> coveted, longed for, outraged and rejected at the same time, the city is actually a huge object of desire. The city spectacularises everyday life, giving visual meaning to the world of people, things and exchanges. A city of encounters and mismatches. Different gazes construct images and representations in infinite compositions. At different scales

[15]On this concept of urban landscape, see Cavalcanti (2001, p. 14).

> and levels, it also allows disparate activities to be concentrated at spatially localised points, revealing worlds near and far. It brings together and disseminates culture and knowledge, it lays bare and reveals secrets. The city establishes itself as a privileged space (SILVA, 1997, p.85).

The challenge of understanding the city of Jataí leads to the exercise of seeing and seeing and thus building a representation that takes into account the analysis in progress.

The exercise begins in the early 1980s, when there was a savannah full of its characteristic fruits, such as pequi and gabiroba. With the transformations that took place, at the end of the same decade a housing estate was built to cater for the population growth produced by the new economic dynamic. In one of its squares there is a snack bar selling soft drinks produced by a multinational company. This situation is in line with what Martine (1994, p.8) said about the spatial redistribution of the population in the 80s. The author explains that "[...] changes in the spatial distribution of the population reflect transformations in the very structure of society".

In Jataí, as in so many other places, the cerrado is losing its space to meet the demands produced by "modern" society (which "appropriates" even nature) and its own dynamics, including population. This dynamic is no longer predominantly rural-urban, as it was from the 1940s to the 1970s, but urban-urban (Santos, 1998).

Revealing a dimension of the space produced, the urban landscape of Jataí shows the overlapping of different times. Past elements that resist the speed of today's society provide important information for understanding the city and the relationships it produces.

Urban society, urban life, perceived primarily in the landscape, does not dispense with morphology. The urban is based and realised in the city, a concrete entity (LEFEBVRE, 1991). In the city, in its structure, revealed in the landscape, the urban goes from a mere possibility to the real, embedded in forms, latent, full of answers, waiting to be questioned.

The exercise was to seek a differentiated vision, seeing the parts and understanding them within the whole. Deconstructing the immediate image and trying to understand its compositional process. Seeing through the landscape and seeing far beyond it. Seeing the reproduction of history, the various conceptions of living, inhabiting, working, eating and drinking. The most varied conceptions of living in the city (CARLOS, 1994).

The landscape holds stories, such as the one recorded by Binicheski (2001, p.103) in an interview with a truck driver, a migrant from Rio Grande do Sul who lives in Jataí. He revealed his dream of owning his "little house": "Eight years ago I earned a lot of money, in the first year I bought my little house, here in Vila Paraíso [...]". The neighbourhood where this property is located is the first funded housing development to be set up in Jataí in the 1980s, marking a change in the urban landscape.

This, and so much other information, is recorded in the urban landscape. The place where this migrant's dream came true is Vila Paraíso, the city's first housing estate. Set up in the 1970s, this neighbourhood changed the city's landscape in a context of economic changes, linked above all to the municipality's productive structure.

The act of observing and analysing the landscape allows for a better and broader understanding of society's needs through what has been crystallised. It also allows us to imagine what could have been built by people who were unable to do so. Perhaps there wouldn't have been so many voids in the city, and others in the individuals who didn't have the full right to the city in which they live.

In this sense, the landscape is also human. Inhuman, perhaps. It is an expression of society, also showing its contradictions. It is a constructed space, but it also harbours the movement of life (CARLOS, 1994).

As we move towards a more solid interpretation of Jataí's urban space, the exercise we propose is to (re)see the city. This action consisted of travelling through the city along routes that allowed for a broader and more differentiated view. Each angle allowed for different images, going beyond the limits of form (SILVA, 1997). According to this proposal, the route shown in Figure 47 was drawn up and seven decisive points for analysis were determined.

Looking at the city "from the outside in" takes the observer to a position of externality. The individual who lives on the periphery, in the segregated spaces discussed, has a different view of the city. The peripheral view thus offers different angles for interpretation.

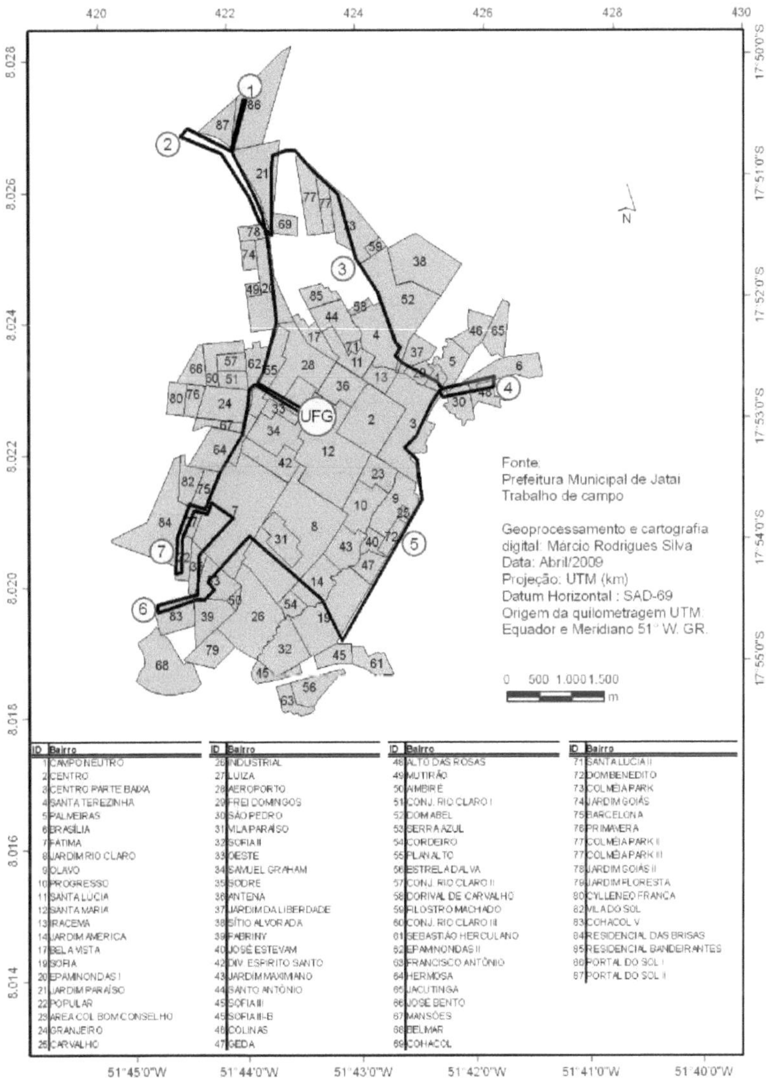

Figure 47: Jataí - Fieldwork itinerary - May/2008

The approximately 40-kilometre route began at the campus of the Federal University of Goiás (Riachuelo Unit). The route began in the north-west of the city (Point 1), leaving an area plotted in the 1970s, passing through others from the 1980s and 1990s and arriving in the current decade. In the same sequence, the route to Point 1 passes through high-, middle- and low-income sectors and arrives at a sector that is likely to contain a higher-income population, considering the land prices charged for the recent sale of plots. As this is a recent subdivision, there is no concrete data available.

Looking towards the city centre in a northwest-central direction, the view obtained is that shown in Figure 48. It shows the most central core at a distance of approximately 5km. A significant distance for

Figure 48: View of the city from Point 1, North-West-Centre Photo by the author - May/2008

the inhabitants of the surrounding neighbourhoods, who don't have adequate urban transport, but who have minimal access by car via the avenue that connects this part of the city. However, access slows down during times when there are programmes at the city's agricultural park, located near Point 2.

The predominant building standard (Figure 46) goes from medium/high at the starting point to simple and then precarious in the neighbouring sector to Point 1. Urban voids (Figures 11 and 12) increase in the direction of the relocation, the dynamics of the city

(Figure 22) decrease and welfare (Figure 44) also increases.

In the same region of the city, looking from Point 2 (Figure 49), you can see the BR-158 in the centre, the Portal do Sol sector (Point 1) on the left, surrounded by a cornfield, and on the right, a forest that is becoming more and more isolated by urban occupation.

The sales advertisement for the Portal do Sol allotment (Appendix B) highlights the presence of the infrastructures that are compulsory when setting up any type of allotment. "Allotment with paved streets, kerbs, electricity, public lighting and drinking water" (Annex B - Allotment publicity material). Since the city still has several areas without such public facilities, the basics become an accessory.

Figure 49: View from Point 2, BR-158 - urban perimeter

Photo by the author - May/2008

Other attractions for the buyer follow: "An innovative development. Allotment with leisure area, square, playground, gym area, jogging track, multi-sports court" (Appendix B - Allotment publicity material). It's hard to see the innovation. The pamphlet continues: "Jataí deserves a place like this"; "in the most valued region of Jataí"; "Grow, grow, for Jataí to grow!".

Of course, the slogans are debatable, as are the parameters for using them. What is not debatable is the simple "growth" of the city by expanding the occupied area, which is not accompanied by the expected development.

Moving towards Point 3, towards the northern end of the city, we have the Colméia Park Sector. The sector has only

There are two access routes, one through the city centre on paved roads and the other on a dirt track, which is the quickest way to get to the west and north-west of the city. The shortest access to the city centre is approximately 4.5km away, leaving from a neighbourhood that includes lower classes, with an average income of two minimum wages. The route to the centre passes through sectors that maintain similar characteristics.

Figure 50 shows the view from the exit of the Colméia Park sector, in the direction of Point 3 - Centre. Temporally, in the same direction, there is a shift from the 2000s to the first half of the 20th century (Figure 9), in the more central neighbourhoods. The route reveals, at the same time, a reduction in welfare payments (Figure 44), an increase in income levels (Figure 42), the dynamics of the neighbourhoods, and an increase in the number of people living in the neighbourhoods.

Figure 50: View of the city from Point 3, north-central direction.

Photo by the author - May/2008

urban (Figure 22), land use (Figures 11 and 12) and building standards (Figure 46).

An interesting contrast in the urban landscape was identified at the next stop. On the way to Point 4, we travelled through sectors established until the 1980s, arriving at the Brasília sector, which in the 21st century retain traces of the rural context of Jataí's gestation, as can be seen in Figure 51. The crystallisation of the rural within the urban setting raises discussions about the fine line between these spaces. Urban segregation at this point takes the individual back to the countryside, which insists on remaining not only in concrete form, but above all through local habits. Although far from places of leisure, the periphery holds

Figure 51: Point 4-0 rural within urban - Brasília Sector.

Photo by the author - May/2008

a certain closeness between its residents. Talking to neighbours and contemplating the street and the surrounding area are ways of bringing people together who are equals in the same neighbourhood, but who are different from those fixed on the distant landscape seen in Figure 52.

Figure 52: View of the city from Point 4, North-East-Centre Photo by the author - May/2008

Approximately 3 km from the centre, the north-eastern region, like most of Jataí's outskirts, concentrates lower incomes (Figure 42), which can be seen in the predominant building standards (Figure 46). Welfare (Figure 44) is also marked, especially in the Jacutinga sector, which is densely populated. Due to its relative proximity to the centre, there is a greater urban dynamic (Figure 22), which has been growing since the 1980s.

Similarly, there is the eastern region (Point 5), from which the city can be seen, as shown in Figure 53, in an east-centre direction. Its peculiarity is the eastern boundary with the Brazilian Army's military area, which did not allow it to expand eastwards.

Figure 53: View of the city from Point 5, east-centre direction. Photo by the author - May/2008

The move from Point 5 to Point 6 took place on the edge of very distinct areas. To the left is the region with the highest rate of emptiness, which decreases towards the centre. Point 6 (Setor Cohacol 5) is home to one of the city's newest housing estates , which consolidates the trend towards valorisation of the intermediate areas. Located on the southern edge of the complex

The view looking out from the city is characterised by rural features (Figure 54). Like other sectors of the edge, there is the presence of agriculture to the west of this area.

The city centre is approximately 5 km away, with a predominantly low occupancy rate in the first 2 km.

Figure 54: Point 6 - Cohacol Sector 5.

Photo by the author - May/2008

areas established mainly between the 1960s and 1980s. Average indicators in terms of income (Figure 42), welfare (Figure 44) and urban dynamics (Figure 22) predominate in this region, minimised by the large size of the area. The most critical indices are the low standard of construction

(Figure 46) and the urban voids (Figures 11 and 12) worsening towards the centre-south.

The route to Point 7 passes through Vila Fátima (built in the 1960s), on a stretch that is out of keeping with the characteristics of the south-western region. Figure 42 clearly shows the presence of high-income levels, clearly perceived by the neighbourhood's construction standards. At Point 7, there is another housing estate opened in the current decade, which is being publicised and sold. In Figure 55, the view in the foreground is of the Brisas Residential sector (the most recent), followed by more vacant areas opened in the 1990s and 2000s.

The Residencial das Brisas sector (Annex C) and its neighbour, the Morada do Sol sector (Annex D), use the

Figure 55: View of the city from Point 7, south-west to centre. Photo by the author - May/2008

proximity to the Condomínio Residencial Barcelona, perhaps to give potential residents the illusion that living near an area with higher income levels will bring them closer to this standard. However, the wall creates segregation, not just physically, but mainly socially. The "dream" of living close to the "elite" can be acquired "in small instalments", which is already different from the neighbours on the other side of the wall. The region described here is characterised by high levels of emptiness (Figures 11 and 12), low urban dynamics (Figure 22), average incomes (Figure 42) and high levels of welfare (Figure 44) concentrated in specific sectors. The construction standard (Figure 46) is simple and precarious, tending to increase in newly opened areas due to the price of land.

On the way back to the starting point is, still in the south-west, the Barcelona Residential Condominium, demonstrating self-segregation. Set up at the end of the 1990s, it has the peculiarity of many vacant plots (approximately 30 per cent of the plots are occupied), demonstrating that acquiring them did not mean living close to "equals" in the first instance, and further research is needed in this regard.

At the end of the route, the vision of the city was worked out. This is another element for the debate.

CHAPTER 4

ACCESS TO URBAN LAND: THE CONSOLIDATION SEGREGATED AREAS

"They run like mighty men; like men of war, they climb walls; and each one goes his own way and does not deviate from his ranks. They don't push one another; each goes his own way; they throw themselves against spears and don't stop in their tracks. They storm the city, they run up the walls, they climb into the houses; through the windows they enter like thieves.

Joel 2:7-9

The last element that is proposed for understanding the process of socio-spatial segregation in Jataí is the value of urban land. This element is significant because its essence lies in the logic of the capitalist system. This system dictates the rules, through its individual or collective agents, for the construction of urban space.

Before presenting the data specific to this discussion, it is worth highlighting the scenario in which they are analysed.

In 2004, the city of Jataí had a total of approximately 37,000 urban plots (JATAI-GO, 2004b). Of these, around 13,000 were unoccupied, or 35 per cent of the total. Bearing in mind that the allotments set up between 1990 and 2004 totalled approximately 5,700 plots, it can be concluded that these were unnecessary to meet the demand for urban plots. Discounting the number of vacant plots from the total number of plots created after 1990, we get a figure of 7,300 plots. In other words, at the beginning of the 1990s there were approximately 7,300 vacant plots in Jataí, a much higher number than the plots that were built on.

If there were no new allotments, the number of vacant plots in 2004 would probably be around 1,600. As a result, the urban space would be denser and there would be fewer vacant plots, moving towards full occupation of the areas that have already been allotted.

Unfortunately, up-to-date data on urban voids has not been made available by the city council. Within the context of the city's urban sprawl, it is thought that this data would reveal an even more drastic situation, ratifying the urban segregation and property speculation identified.

It's important to emphasise that measures to guarantee accessibility to urban land for the entire population must not be overlooked. Considering the hypothesis of not opening new allotments, it becomes necessary to take measures to guarantee a fair price for existing plots which, without new areas to compete in the property market, would tend to rise in price.

A situation like this occurs along capitalist lines. The implementation of new allotments is part of a reality that goes beyond the need for new plots to meet existing demand. Urban land is not available to everyone; people on lower incomes are only able to buy a property in places further away from the city centre and, consequently, from urban amenities.

The power of land ownership in the urbanisation process is considerable. The formal nature of ownership, of the right to property, allows the owner to extract income from the land, even without exploiting it, without touching it, even in an absent manner, which, among other factors, is linked to location and infrastructure (LEFEBVRE, 1999).

According to Kowarick, apud Souza (1994, p.172), when discussing the speculative process in the city of São Paulo: "an allotment is never built immediately after the last one, always leaving an intermediate area that incorporates all the urban infrastructure that had to pass through it, increasing its value". Drawing an analogy to this analysis, while respecting the differences between a metropolis like São Paulo and a city in the interior of Goiás, the reality for Jataí is very similar.

Property speculation is a way of realising surplus value. The *money, commodity, money* circuit *is* completed to the extent that "fattened" land returns more money to its owner. Thus, land becomes a "'good' that can be commercialised, dependent on exchange value and speculation, not on use and use value" (LEFEBVRE, 1999, p.161). "Use value corresponds to need, expectation, desirability. Exchange value corresponds to the relation of this thing to other things, to all objects and to all things, in the 'world of merchandise'" (LEFEBVRE, 1999, p.135).

This context outlines the socio-spatial situation of each inhabitant of the city. One's view of the city stems from one's own situation. On the one hand, the "beautiful city", for those who can afford it, and on the other, the "ugly city", for those who have been displaced by society to the "ugly" parts of the city. The city therefore has several faces. As many as the different ways its residents look at their homes, not just their houses, but the space they occupy in their houses, flats, neighbourhoods, cities and society.

In this respect,

> spatial production expresses the contradictions of today's society in the juxtaposition of wealth and poverty, splendour and ugliness; ultimately, in the segregation stamped on the landscape, and which has its nature in the mode of exploitation of labour by capital based on property, in which men are distinguished by having, that is, by their status as owners of goods (CARLOS, 1994, p.15).

A city becomes good or bad according to whether it fulfils the needs of its inhabitants. But which needs? There are inhabitants with different incomes and, consequently, "different needs". Meeting everyone's needs is a difficult task in a reality that doesn't cater for basic needs. In this way, ensuring that residents are well connected to their places of residence, work, education, consumption and leisure is an important measure in determining the quality of urban life.

Jataí's urban dynamics are greatest in the central areas, and the urban zoning itself perpetuates this situation. There is no direction, no ramification, so that the trade and service zones (Structural Zones) serve the city as a whole. The result: there is no good articulation between the areas. Since the space is fragmented, it is essential to promote its consolidation as a living city, with its own dynamics, the dynamics of the city.

If segregation, in the sense of living apart, places individuals closer together, in places that allow them to be grouped together with a good quality of life, it's safe to say that it's not all bad. Segregated spaces group together equals, above all by income level.

Would it be bad to live close to your peers? A good quality of life, with facilities that make it possible to connect with the city, in a process that meets the needs of the group, would allow for a closeness that has already been lost in many large centres, but which can be recovered in a city like Jataí. Even in a process of participation in neighbourhood and community management, taking advantage of the city's size, promoting a better quality of life, along the same lines as those proposed by UNESCO. Jataí has the potential to be a human settlement that can offer a better quality of life to the resident population, due to its urban scale, and that can allow citizens to participate more in the government and management of the city (UNESCO, 1999).

Santos (2000) also raises the issue of the right to the city, the right to one's surroundings, in a reflection on the privatised spaces in the city, public spaces typical of traditional urban life.

We have to buy what is public. Leisure in the city is paid for, inserting the population into the world of consumption. "Those who can't afford the stadium, the swimming pool, the mountains, the fresh air, the water, are excluded from enjoying these goods, which should be public, because they are essential (SANTOS, 2000, p.48). As living on the outskirts is, in most Brazilian cities, the fate of the poor, they are condemned to not having social services or to using them precariously, even if they pay extortionate prices for them (SANTOS, 2000, p.47).

In a correlation between the location of people and their social and income level, urban space is occupied differently according to the classes into which urban society is divided (SANTOS, 2000). Thus,

> Every man is worth something because of where he is: his value as a producer, consumer and citizen depends on his location in the territory. Their value is constantly changing, for better or for worse, depending on the differences in accessibility (time, frequency, price), regardless of their condition. People with the same potential, the same education and even the same salary have a different value depending on where they live: the opportunities are not the same. That's why the possibility of being more or less of a citizen depends to a large extent on where you are in the territory. While one place becomes a condition of their poverty, another place could, at the same moment in history, facilitate access to those goods and services that are theoretically their due, but which in fact they lack (SANTOS, 2000, p.81) (Emphasis in original).

The following passage is a good example of the situation of the forgotten periphery. The neighbourhood mentioned in a local newspaper report is located in a peripheral area with low urban dynamics.

> Unfortunately in Jataí it is still a reality to find neighbourhoods abandoned by the public administration, proof of this is the precarious situation in which the residents of Jardim Floresta live [...] The residents

say they are already tired of looking for the city hall to solve the neighbourhood's problems [...] the city hall says that this neighbourhood doesn't exist on the map [....] which they think is absurd, they can't even get a telephone line there because the operator looks for the neighbourhood and can't find it, when they need to order some medicine or even a snack, they can't get anything, because when they tell them the address nobody wants to deliver it, not even motorbike taxi drivers take passengers to the neighbourhood. They go as far as Av. Goiás [the nearest paved road] and walk the rest of the way.

"We residents of Jardim Floresta are in a situation of total abandonment by the government. A machine was driven through here, but it hasn't solved our problem, in my opinion it's made it worse because it's only increased the dust. Before we suffered from the number of holes in the street, now it's the dust. My daughter has allergic bronchitis and can't even stay at home, I've had to take her to my mum's because it's impossible for her to stay here. After we complained a lot, the rubbish collection started, but that's not enough because we pay our taxes. We've been told by the town hall that our neighbourhood practically doesn't exist, that it's not on the map, and this was also said by the current administration on a radio station for everyone to hear, if our neighbourhood doesn't exist, how come the IPTU comes every year for us to pay." (CARLÚCIO ASSIS, 2009).

Considering that there are separate spaces, separate groups, and that the spatial production of the city creates "rich" and "poor" spaces, it is necessary to give everyone the necessary conditions for the full development of individuals. The reality described needs to change.

So who will pay the price for fragmentation? The fragmented city is more expensive, from a simple spatial view of infrastructure displacement. We only have to remember the urban voids, the areas purposely "forgotten" in the middle of the road and which are growing at high speed.

To discuss socio-spatial segregation in Jataí is to act to change a situation that can still be changed at little cost. To ignore its existence is to assume a future with not very different, and possibly much worse, realities.

The increase in the price of urban land and its appreciation is also due to the actions of public authorities.

> The location of infrastructure and collective use facilities under its responsibility, zoning laws, plans for the "urban revitalisation" of certain areas of the city, the provision of financing for the purchase of housing, and the production of housing by the public authorities are some of the examples of how the state plays an active role in the formation of urban land rent [...] (BOTELHO, 2007, p.78) (emphasis added)

Urban land rent has a lasting impact on urban development, and its main spatial manifestation lies in the phenomenon of segregation, produced by the mechanisms of land price formation (Lojkine, 1997). At times, one can see similar action by theoretically different spheres: State and Company.

> The state and business [...] endeavour to absorb the city, to suppress it as such. The state acts above all and the company below [...]. The state and business, despite their differences and sometimes their conflicts, converge towards segregation (LEFEBVRE, 1991, p.95).

The search for profit through property speculation makes urban property increasingly unaffordable for the lower income classes. With areas practically "closed" to the public and the need to meet the demand resulting from local population growth, the new areas set up bring with them the valorisation of intermediate areas. Properties with a privileged location, closer to urban amenities and the main economic activities, provided their owners with an even higher income.

This rent, land rent,

> has its origins in modes of production that predate capitalism. However, it finds its place in the social articulations of this mode of production, and is even part of its process of reproducing relations of production. The hierarchisation of different places according to the income they provide (or the price charged for land) and the exclusion of the great mass of the population from access to land, i.e. the reproduction of non-owners, are conditions and results of the specifically capitalist process of reproduction. Furthermore, land ownership and rent (the payment for its use by a third party) are a legal and ideological foundation of the capitalist economic-social

> formation (BOTELHO, 2007, p.67).

Land rent "[...] is a fundamental element for understanding the hierarchisation of urban land use, for the accumulation of capital and for the reproduction of capitalist relations of production, as well as being an important instrument of socio-spatial segregation" (BOTELHO, 2007, p.69).

Through the previous analyses, especially those of income levels, it is possible to make a correlation between the distribution of individuals according to income and the price of urban land in Jataí. According to Ribeiro (2005, p.89) "[...] property prices have become the central mechanism for distributing the population in the city's territory". This means the most direct translation into the organisation of urban space of the income inequalities that exist in society.

The price of "living in the city" thus co-operates with the consolidation of segregation. Whether it's optional, in the case of self-segregation, or forced, reality points to Lefebvre's observations:

> Despite good humanist intentions and good philosophical will, practice is moving in the direction of segregation. Why? For theoretical reasons and because of social and political causes. On a theoretical level, analytical thinking separates and decouples. It fails when it tries to synthesise. Socially and politically, class strategies (unconscious or conscious) are aimed at segregation (LEFEBVRE, 1991, p.94).

The price of urban land, and the income acquired through it, are also obstacles to the social ownership of one of the most important means of production: land (BOTELHO, 2007); thus contributing to the continued reproduction of the current social condition.

Lojkine (1997, p.189), when dealing with the issue of land rent, distinguishes between three main types of urban segregation:

> 1) An opposition between the centre, where the price of land is the highest, and the outskirts [...];

> 2) A growing separation between the areas and housing reserved for the more privileged social strata and the popular housing areas;
>
> 3) A generalised breakdown of "urban functions", disseminated into geographically distinct and increasingly specialised zones: office zones, industrial zones, housing zones, etc. This is what urban policy has systematised and rationalised under the name of zoning.

The third phenomenon is directly linked to land income when the mechanism of social selection, constituted by the growing difference between land prices in the periphery and those in the centre, places the most "noble" areas in the hands of those with the highest incomes. Zoning formalises the market game. The land rent game excludes those who don't pay its price. The excluded are identified in their own zones.

In order to get as close as possible to the reality of the city, we used the average sales values of plots and buildings, calculated on the basis of information from the local town hall for the year 2004, to determine the price of urban land. As no data was available for a more recent period, it was decided to make a projection of the 2004 values based on the NATIONAL CONSUMER PRICE INDEX - INPC[16] , accumulated over the period. Thus, the figures presented for 2009 consist of the original figures corrected by 25.98% (IBGE, 2009).

Based on this data, it is possible to visualise the past reality and associate it with the distribution of the population in the city, as well as to project a scenario that is close to the current reality, with more up-to-date prices.

In general, we have an urban design that fits in with that outlined by the distribution of the population according to income, with land prices rising in the more central areas. This behaviour can be

[16]The INPC was chosen because it is the most widely used index in the market for readjusting completed properties or rents.

observed, in a didactic way, in the five areas used for the spatialisation of income levels in the city.

In order to avoid distortions in the analyses, the industrial sector was excluded (Figure 37 - neighbourhood number 26), which has very different characteristics from the other areas of the city, especially in terms of the value of the buildings, since it includes large areas and high-value industrial buildings, compared to the residential standard.Figures 56 to 59 show the average sales values of plots and buildings. This spatialisation allows for the following analyses

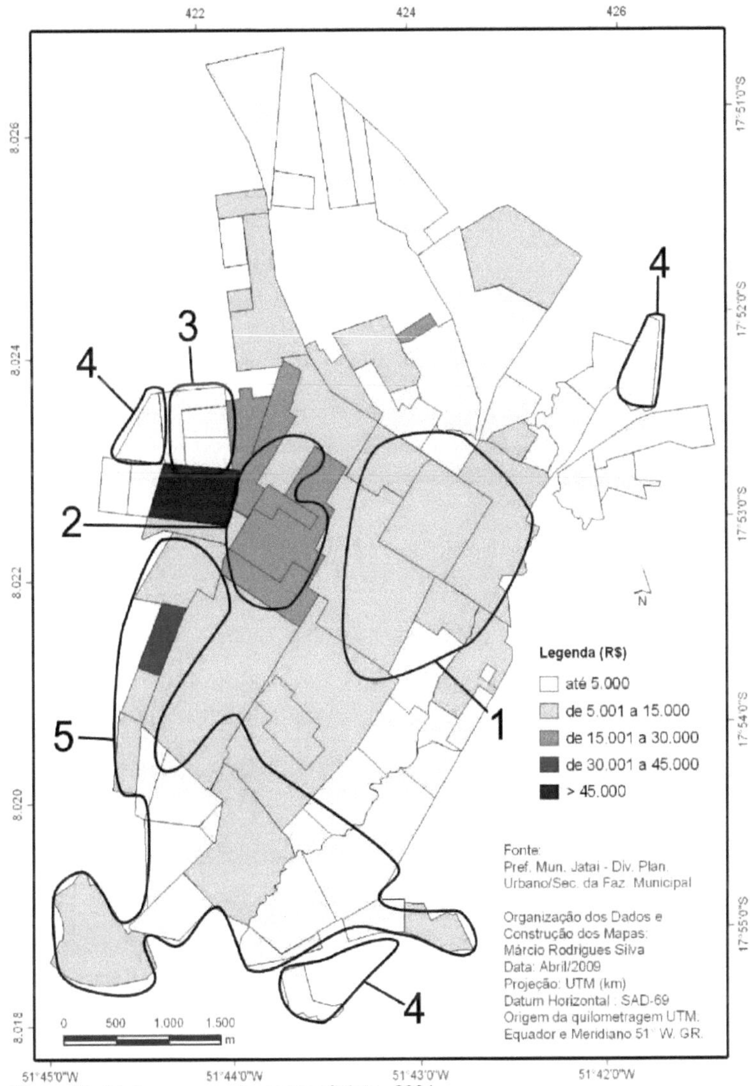

Figure 56: Jataí: average sales value of plots - 2004

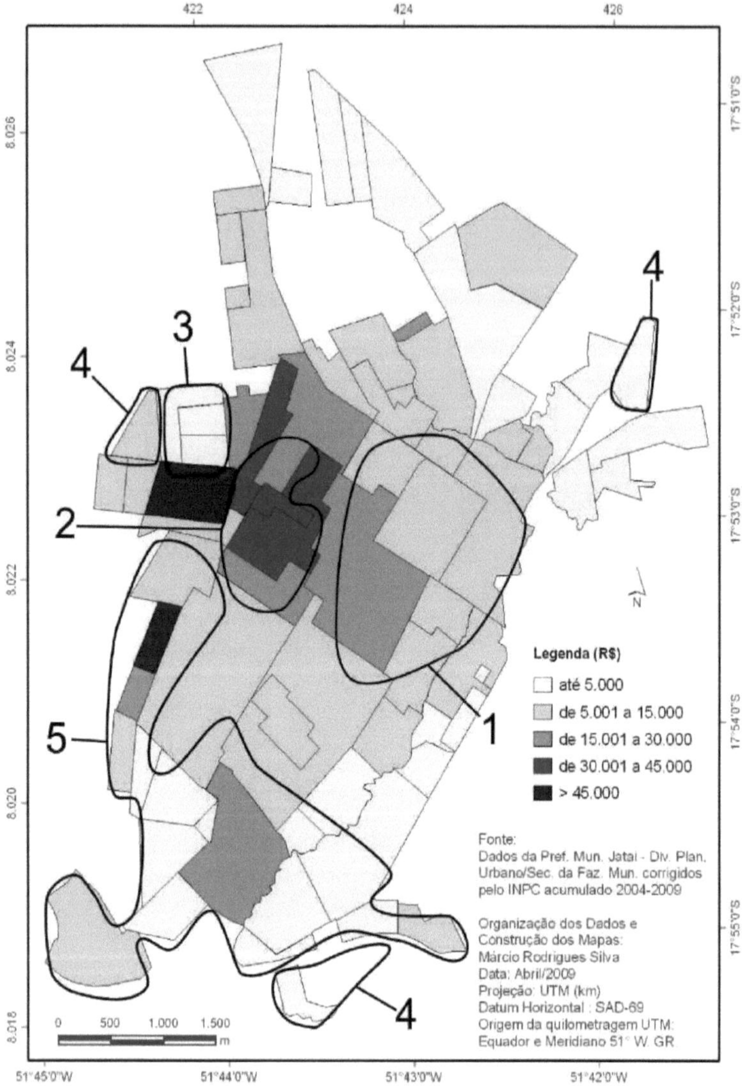

Figure 57: Jataí: average sales value of plots - 2009

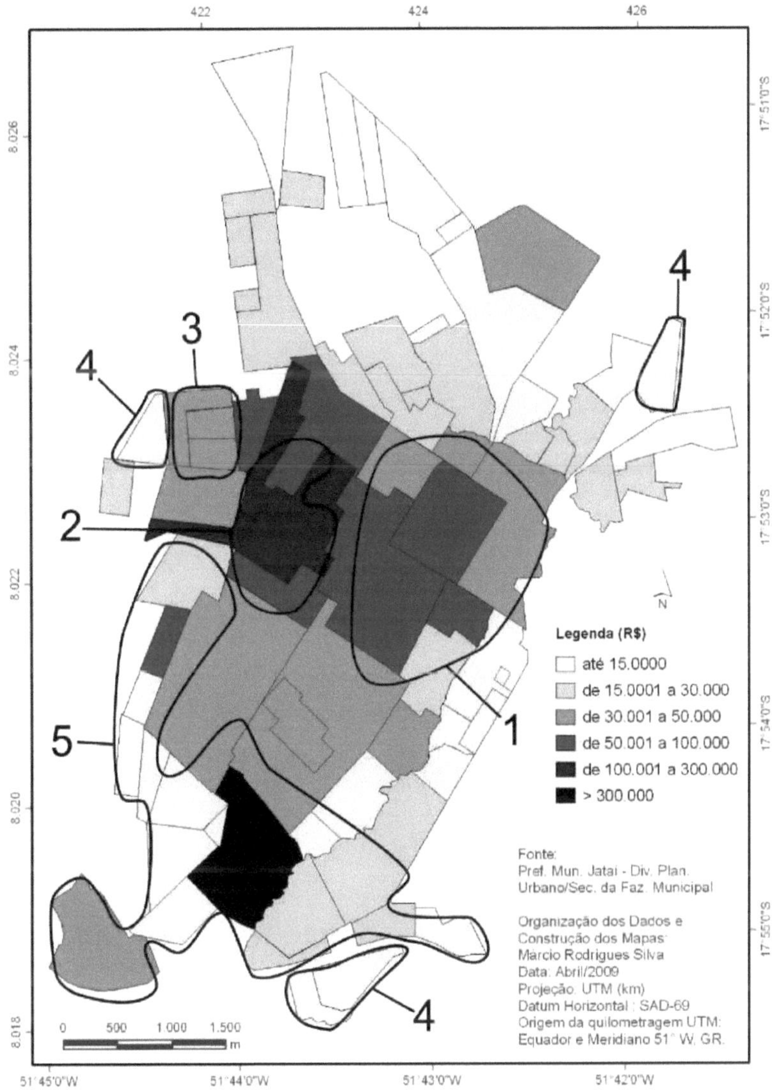

Figure 58: Jataí: average value of buildings - 2004

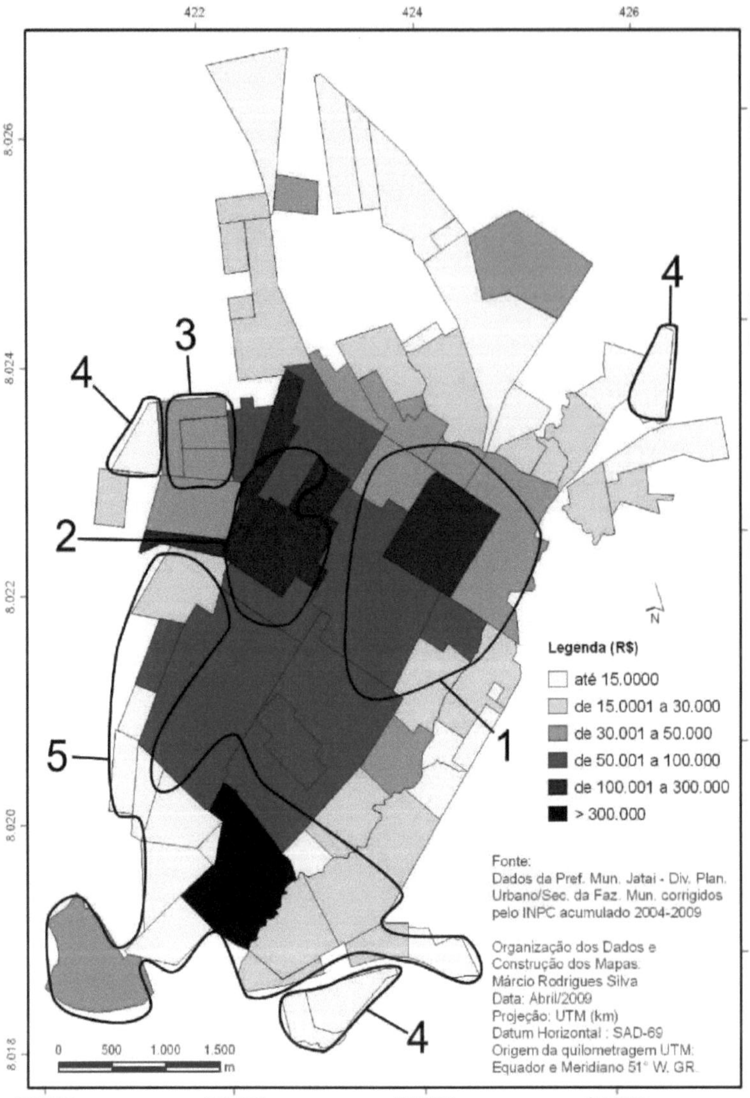

Figure 59: Jataí: average value of buildings - 2009

- <u>Area 1: Original city centre</u>

In the central area, which contains the population with the highest incomes, there are the highest building values.

For the values of the plots located in the lower classes, the reading must be different, since there are few vacant plots in this area. The lower values probably refer to plots with smaller areas and poor location for the predominant commercial characteristic of this centre.

In short, this is an expensive area. High values predominate for standard, built-up properties.

- Area 2: central-western part of the city - upmarket neighbourhoods

The same pattern can be seen in all four maps: high urban land values. Where the highest incomes are located, this pattern prevails for both plots and buildings.

- Area 3: Middle-income housing estate

The analysis of this area basically takes into account the value of the buildings. As this is a housing estate, the vacant plots are public areas.

Building values are in the middle range, both for the 2004 values and for the 2009 projection, at the same level as the income found in this area.

- Area 4: Low-income housing estates

In the sectors highlighted, which correspond to low-income housing estates, the correlation *between land prices and income levels* is corroborated, with lower incomes acquiring compatible properties. Buildings and plots in these areas are in the lower price brackets.

- Area 5: Area with large urban voids - over 50%

    The plots in this area are in the lower price brackets, which would lead to greater densification, considering only the price of land. However, it should be borne in mind that the densification of the more peripheral areas was directed by the donation of houses/plots, which directly influenced the urban emptiness of area 5 and the "fattening" of this land.

    With a high rate of urban voids, the lack of infrastructure in these areas does not attract inhabitants. The price of buildings in the higher range is distorted by the average, since there are high-end buildings concentrated in a small part of the area.

This correlation between land prices and income levels in the selected areas leads to a similar reading for the other areas of the city.

Analysing land prices completes the picture needed for the final considerations of this work. The pattern is always the same, revolving around the income of the city's inhabitants. This pattern makes socio-spatial segregation in Jataí clear.

City dwellers, victims of segregation, residents of "poor areas", are part of the political action of segregation, which hides the problem of the city and the urban from them (LEFEBVRE, 1991). Unaware of the urban, they live *in the city,* but they don't live *the city.* They are in it, but they are not part of it. More than inhabiting it, we need to live it.

> The right to the city manifests itself as a higher form of rights: the right to freedom, to individualisation in socialisation, to habitat and to dwell. The right to *work* (to participate in activity) and the right to *appropriation* (quite distinct from the right to property) are implicated in the right

> to the city (LEFEBVRE, 1991, p.135) (Emphasis in original).

Actions, both theoretical and practical, must converge so that everyone has the right to the city. The right not just to the technical, concrete and cold city, but to the human city, with its warmth, smells and flavours. To this end,

> urban development policies should seek out the conditions of urbanity, with the aim of redeeming collective spaces as a sign of the new city. Rescuing the street not only as a means of circulation, but also as a place for people, everyday life and the exchange not only of goods... but also, why not, of love? In addition, they should pursue the search for a time for encounters that goes beyond the encounter for the exchange of things. We need to create times and spaces for life in all its dimensions. This involves recovering citizenship, which in the city requires the realisation of a decent life, which presupposes access to dignified conditions of survival. Although the absence of goods and services is abominable, the lack of time, leisure and information, as well as natural goods such as clean air, water, light and trees, are equally abominable. Thus, the rescue of citizenship contains the dimension of the conditions necessary for the reproduction of life in its broadest sense (OLIVEIRA, 1997, p.270).

4.1 - FROM URBAN PROBLEMS TO THE IMPLEMENTATION OF SOLUTIONS

Identifying urban problems doesn't lead to a solution to the problems of cities. It's actually the starting point so that, based on the diagnosis, new paths towards a solution can be mapped out.

Now that the city of Jataí is understood, or at least known more intimately, through "looking at it and talking to it", there is no room for negligence in not implementing solutions. And what is the role of the researcher in this movement? It is understood that it is to promote debate and raise questions that lead to the improvement of the city. In this sense,

> [...] it is necessary to answer a series of questions that affect the daily lives of people in large, medium and small cities. The "answers" must take into account specificities, a

> dimension in which they may not even be that important, but above all they must reflect an understanding of the urban social process, which necessarily includes an understanding of the totality (OLIVEIRA, 1997, p.270).

Along these lines, the action proposed here is more about debating the issues than "discovering the problems". Weren't they already known? We think they have been known for a long time. At the very least, since the beginning of the 2000s, the city has been known and management proposals have been translated into the urban masterplan.

It is understood that solutions exist, but in many cases they are not implemented. "The real or imagined diagnoses are almost always correct. However, the practicality of the proposed actions falls short of the recommended solutions" (OLIVEIRA, 1997, p.268). Therefore, it is not necessary to propose new paths. The existing paths will be sufficient as they are put into practice. New proposals are left for a post-concretisation context.

Let no injustice be done. It is true that there are actions that have been implemented and have resulted in improvements for the city. We can quickly recall: the urbanised areas that have brought life to regions that used to keep city dwellers away; the housing estates (those built in accordance with the law) that have offered decent housing to the population; the museums, libraries and cultural centres, so rare in inland cities.

We can therefore see that it is possible to build a more harmonious urban whole. We can therefore dare to say that the solution to the problems encountered lies directly in a question of scale and continuity.

The scale needs to be expanded. Existing interventions are piecemeal, catering only for the immediate surroundings. Worrying about the expansion of public services, urbanisation and all the legal guidelines that have already been drawn up, on a scale that involves the whole city, is fundamental if growth that doesn't stop is to result in development.

This directly involves the continuity of public actions. What we realise is that changes in the municipal administration have a direct impact on urban actions, which should not be "political". Every four years the city remains uncertain, and it's not uncommon for public works to be left aside due to divergent interests on the part of one group or another. Generally, with each new government, the first year is spent "redoing" what has already been done, giving the city the "face of the current administration". The second year is for planning. The third year is for implementing the planned actions, which stop at the end of the fourth year when "politics" prepares for another cycle.

The central question is then to put what is notorious into practice. As Souza (2003) states, the focus should not be on finding a *scapegoat*, but on discerning and tackling the causes of urban problems.

What should be tackled? The structures of power, the channels for distributing wealth and everything that is vitiated and generates social injustice (SOUZA, 2003). We have clearly seen that income distribution is directly linked to access to urban land, limiting the right to the city. So this must be addressed.

And what about "interest groups"? The only group that matters is the one that involves all citizens. There should be no room for groups that own urban land and regulate access to it. For those who segregate themselves, by choice, and generate the segregation of others, by imposition.

These situations, which aggravate the city's problems

(SOUZA, 2003), are not the fundamental problems. Fundamentally, we need to implement what already exists. There are limitations, of course. Budgetary, institutional and other constraints. But we understand that, within these limitations, it is possible to improve the city.

CHAPTER 5

ANALYSING THE RESEARCH PROCESS

The wise hoard knowledge, but the mouth of the foolish is an imminent destruction.
Proverbs 10:14

At the end of yet another stage in the study of the urban space of Jataí-Go, part of the "scientific curiosity" that motivated the research has been satisfied. Yes, this is only the conclusion of one more stage, since the object of study is not considered to have been exhausted by the geographer's gaze. On the contrary, as answers emerge, new questions follow, in a movement that resembles a geometric progression and at the same time moves away from it because it is not so exact.

It is in this dialectical relationship that some fruits are harvested. Fruits that lead to an answer to the initial question: is there socio-spatial segregation in Jataí?

By analysing its internal structure, in terms of socio-spatial distribution in the urban arrangement, the aim was to contribute to the process of "unveiling the city".

Jataí's urban space is not shaped in isolation, but in the context of its micro-region, the state, the country and even the world, as discussed above. In this context

he answer to the question raised lies in the following points.

To get to know Jatai's urban space, it's not enough just to observe it in its raw form, as it is visibly presented. Understanding the formation of this landscape, which brings with it encounters and mismatches, is a task that involves interpreting what is shown in the foreground. It was therefore necessary to unveil this space. It is in this activity that the eye of the researcher, the scientist, must be alert to the fact that what is visible is not always what is real. This analysis reveals a city full of contrasts, most of which are determined by a social reality embedded in today's unjust capitalist context. Social differences that when grouped together form an unequal urban space.

This inequality is not formed over a short period of time, it starts at the beginning of the municipality's history, it starts with the gestation of the large farms formed in the region for the development of agriculture and is directly linked to the current economic context and modern agriculture. The latter has more markedly imprinted today's urban characteristics. The urban agglomeration has an intimate relationship with the implementation of modern farming in the municipality and its surroundings, which brings development to the region but, on the other hand, leaves side effects such as the swelling of urban centres and inequality in income distribution.

In the origins of the municipality, which date back to 1836, when the first settlers arrived in the region, the current reality could not have been imagined. Traditional farming activities were not enough fuel to power the dynamic engine that currently drives the region. A transformation was needed, which occurred in a more contemporary period. It was above all from the 1970s onwards that significant changes in the use of the region's land could be seen.

As a result of these changes, the municipality will experience significant growth, with grain production for export standing out, with a dependence on the global being reflected locally. Also, as a result of the changes that have taken place, there is a need to control the occupation of urban land, so that the more than 90 per cent of the municipality's population living in urban areas does not become an additional problem for the public administration.

It is in this dynamic that Jataí has seen a huge growth in its urban area with the implementation of various allotments, most of which are based on the capitalist principle of profit. The action of capital is increasing and oppressing the lives of lower-income individuals who are unable to acquire a plot of land in the city.

These actions have consequences for the city, creating a fragmented and expensive space for taxpayers. The urban voids, waiting for a perceived increase in value with each passing day, above all due to the observed process of "expansion of the city centre", displaces a large population contingent to the poor periphery. In this process, public facilities are brought in which, even though in many cases they are inferior in number and quality to local needs, subsidise the appreciation of the intermediate areas (it should be emphasised that these facilities should have been present before the arrival of the inhabitants of these areas).

The urban arrangement in question is therefore contradictory. Contradictory space, contradictory society. Some move to the periphery by imposition, others by choice. Society even differs in the composition of the peripheral areas, since part of the elite has chosen to surround itself with high walls in its gated community, "far away" from the problems of society built by the system that gives them privileged conditions in society.

These are some of the mismatches perceived when analysing the occupation of this part of the Goiás Cerrado. Certainly the original vegetation, with its twisted branches, found in remote times, was more harmonious than the city built after it was felled. The trees of the Cerrado formed a cohesive space expressing a natural unity. The city forms a fragmented space, a space built without the harmony necessary for the fluidity expected in a modern society.

If space is not harmonious, it is dynamic. This can be seen in the movement promoted by the establishment of economic activities, especially since the 1970s. In a city with basically agricultural characteristics, the services sector comes first in terms of the number of establishments set up, in a development that reveals the movement promoted by local wealth and indicates new directions for the city's economy. This dynamism is also due to the consolidation of the urban space which, starting from the original urban core, expands in search of more and better spaces, consolidating the expansion of the city centre and probably supporting the subsequent emergence of a new centre, or even sectors specialising in specific services.

From the arrival of the agricultural frontier to the present day, there has been a major transformation. Today, occupation is organised and directed by capital, or rather by agents of capital, by individuals at the service of capital. The town that depended on the original small centre, around the church square and the mansions, no longer exists. Not that this crystallisation has dissolved in space-time, although part of the architectural memory has been destroyed. What is happening today is a remodelling of the "old" in favour of the "new". Old forms are given a new function: the old mansion is now a museum and serves tourists; in the square it's rare to find neighbours chatting, and periodically there's a fair that brings together mainly young people who listen to non-traditional music.

The changes are not restricted to the centre, but spread throughout the urban fabric, mainly due to the addition of new spaces. The forest, once far from the houses, is now "isolated" by them, bearing witness to the transformation not only of the city, but above all of the society that makes it up.

This is the existing socio-spatial arrangement. Society has become spatialised. It spatialises itself every day. It imposes its marks. Today this process already has a certain direction. It's the Urban Master Plan that, in a way, shapes the new areas, as long as its requirements are met; if they aren't, the urban arrangement is more spontaneous, favouring individuality and damaging the city.

Past mistakes are consolidated as a warning that the way forward will be different, and this depends on taking advantage of the potential offered by society, fundamentally that of scientific knowledge. It is up to public leaders to know how to make use of this potential in order to promote the real development mentioned above.

With so many changes, movements and dynamics typical of the city's growth, the urban space responds to the actions suffered. The high rate of urbanisation is not restricted to numbers. These are individuals who need space in the urban layout. Migrants arrive to put more pressure on the demand for housing, space and public services. The rural population is no longer rural, it congregates in the city and does not become urban, because it retains characteristics, traits and customs that are typical of the countryside and, perhaps, the desire to return to it. All these actors are part of a theatre run by capital. As for the roles they play, we don't know for sure, it will depend on the local and global situation at the time.

The city responds to everyone with answers that don't always please. Or rather, it pleases some and displeases many others. The city responds with urban sprawl, ecological parks, services that are growing every day, free low-income housing, state assistance for unemployed families - a reward for those who have not managed, on their own merits, to survive with dignity.

There are several answers. All of them, however, are

conditioned by higher wills and are generally aimed at serving individual interests in a process of socialising losses and privatising profits. But what is the loss or profit? Of the urban whole, which has been transformed into a profit-making enterprise.

Today, an image of the city is created to be sold externally (Appendix A). However, it doesn't publicise social inequality; it doesn't publicise the situation of the more than 20% of the population found in poverty. It is not publicised that the richest 20% of the population hold around 64% of the local income. Segregation is taken for granted. The urban patchwork quilt is decorated with pieces that stand out for their strong colours, for spaces that are separate from city life. Spaces with individuals who also have the right to the city.

There is no room in the war of places for bad marketing. It's clear that the city has great potential. However, social equalisation must be promoted so that it is not necessary to "make up" the city in order to promote its development. With more social justice and equal conditions for all, the crystallisation of this aspired society will take place in the urban space. The "ideal city" will thus be the result of fundamentally social actions. It will reflect the valorisation of the majority of society that is currently so forgotten.

Improvements in local urban conditions are linked to the city's economic development, since the two are inextricably linked. The decades between 1970 and 1990 saw significant economic growth and, as a result, greater dynamism in the city as a whole. During this period, particularly in the 1980s and 1990s, the urban space began to be structured more consistently, either by consolidating an economic centre with an emphasis on the service sector, or by occupying intermediate areas between the centre and the periphery. Encouraging the local economy, but with an eye to ensuring that only the positive effects are retained, is crucial for the consolidation of Jataí at local and regional level.

By concentrating services, the city attracts a whole range of positive factors for local development. By structuring the infrastructures that are fundamental to the smooth running of urban activities, Jataí will be able to strengthen its role of influence in the south-west region of Goiás. Offering quality services in terms of training - with an education system that meets demand, especially at higher education level; culture and leisure - investing in local tourism and its potential; urban transport - serving the entire urban network; health, safety and others is fundamental in a city that is interested in "competing" in the war of places. Let the competition take place, if necessary, but let it be with a city that is presented in a real way, a city that is equalised and not segregated.

The city has a contradictory spatial distribution. The urban fabric is made up of components that, when joined together, give the city a socio-spatial configuration that reveals the society behind the visible. It ranges from imposing buildings in upscale neighbourhoods, with all the necessary infrastructure for a good quality of life, to inadequate housing located in neighbourhoods for a section of the population that has been excluded from the benefits offered by capital.

Segregation is present. A broader look, involving the whole urban space, reveals contrasts that are initially perceived by the colour of the streets, which lose their "shine" as you move away from the city's prime areas. Those who can't afford the high price of urban land don't just move away from the city centre, they move further and further away from access to culture, leisure, education and health, among other benefits that should be available to everyone.

When we talk about segregation, there is a tendency to imagine that the segregated population is located only on the outskirts of the city. However, the term periphery has lost much of its characteristic in recent times, and this characteristic is very pejorative. Today, the periphery does not define itself. Unfortunately, there have been no changes for the classic inhabitants of the "peripheries"; what is happening is that the elite are moving into these areas, building a "noble periphery" with their private condominiums. And this is not only seen in large centres, it could also be observed in this object of study. Just as new so-called noble areas have been chosen in places far from the city centre, there has been an inappropriate occupation of more central areas.

Another interesting aspect of this analysis refers to the redevelopment of some parts of the city. The so-called ecological parks have given new life to areas that were previously forgotten in the urban landscape. As a result of the creation of these parks, however, there have been some side effects, such as the increase in value of the areas located near them, which has supported property speculation and, consequently, the maintenance of several urban voids. As part of the city's "make-up", these places have become a reference point for the population, especially in terms of leisure. Unfortunately, these are one-off interventions, necessary for so many other places neglected in this process.

It is clear that redevelopment actions are fundamental in the process of structuring and restructuring urban space. However, there is a warning that actions such as these should not become "beautiful problems". The instruments for controlling the use of urban land, such as the City Statute, must be put into practice, thus leading to a fairer process that includes all social spheres and contributes to the harmony of the urban whole.

The transformations that have taken place in Jataí in recent years show that well-targeted public actions can positively alter the urban reality. However, they warn that these actions should be taken consciously and with an appropriate technical and scientific basis in order to serve society and not just a portion of it, promoting the much-dreamed-of "real development" and an improvement in the quality of life of the people of Jataí.

Although the problems are still many and need constant attention, the urban fabric is considered to have evolved. This fact increases the responsibility of those in power in the face of a reality that differs greatly from the past. Today, Jataí has a consolidated urban space. Contemporary society's capacity for transformation is high and fast. Urban management cannot lag behind.

It is possible at this point to present an answer to the initial question: yes, there is socio-spatial segregation in Jataí.

The questions were asked. The spaces were identified and mapped. This work comes to an end with the aim of contributing to solving the problems raised, so that the mistakes identified are not repeated and so that successes become the norm for the city.

The aim of these brief and preliminary observations is not to reach immediate conclusions on a subject that involves so many other variables. The aim is to demonstrate some points that are relevant to the study of Jataí's urban space and its dynamics. Providing subsidies, promoting debates and discussions for a more rational occupation based on principles that lead to a better quality of life for the local population is also the aim of this work.

The studies continue in the hope of contributing scientific analyses and data worked out within the Geographical Science. After all, according to Monteiro (1995, p. 27), "the greatest charm of

'Geography' lies in the complicated weave of human interactions with the Earth". It is therefore up to geographers to investigate how this fabric was woven.

REFERENCES

ABREU, Irlane Gonçalves de. Urban Geography: Questions about its nature and object. In: CARLOS, Ana Fani Alessandri (Org.). Os caminhos da reflexão sobre a cidade e o urbano. São Paulo: Edusp, 1994. p. 129-135.

ENVIRONMENTAL AGENCY OF GOIÁS. GeoGoiás 2002. Brasília: Goiás Environmental Agency: CEBRAC Foundation: UNEP: SEMARH, 2003. 272p.

BARREIRA, C. C. M. A. Vão do Paranã: The Structuring of a Region. Brasília: Ministry of National Integration: Federal University of Goiás, 2002. p. 157-183.

BENKO, G. Economia, Espaço e Globalização na Aurora do Século XXL São Paulo: Hucitec, 1996. p.50-66.

BINICHESKI, MARIA INÊS. Jataí's land structure: migration from the south and its relationship with local development. 2001, 125 p. Master's dissertation - Federal University of Goiás, Faculty of Law, Goiânia, 2001.

BORGES, B. G. O despertar dos dormentes: estudo sobre a estrada de ferro de Goiás e seu papel nas transformações das estruturas regionais: 1909-1922. Goiânia: Cegraf, 1990. 128p.

BOTELHO, Adriano. O urbano em fragmentos: a produção do espaço e da moradia pelas práticas do setor imobiliário. São Paulo: Annablume, Fapesp, 2007. 315 p.

BRANCO, Maria Luisa Castello. Some considerations on the identification of medium-sized cities. In: SPOSITO, Maria Encarnação Beltrão. Medium-sized cities: spaces in transition. São Paulo: Expressão Popular, 2007. p.89- 112. (Geografia em movimento).

BRUMER, A. Family farming in Brazil. In: LAMARCHE, H. (Org.) Family farming. Campinas: UNICAMP, (s. d.), p. 179- 234.

CAMPOS FILHO. Labour and development are part of history. Tribuna municipalista, [S.l.: s.n.], 1 a 15 dejan. 2004. Special supplement-Jataí: New tourism centre in Goiás. 8p.

CARLOS, Ana Fani Alessandri. The (re)production of urban space. São Paulo: Edusp, 1994. 270 p.

CARLÚCIO ASSIS. "Jardim Floresta - a neighbourhood that doesn't exist! Jornal da Cidade, Jataí, p. 1-1. 24 Sep. 2008. Available at:< www.jornaldacidadejatai.com>. Accessed on: 12 June 2009.

CARVALHO, Edemir de. Social exclusion and the growth of Brazil's medium-sized cities. Scripta Nova. Revista electrónica de geografia y ciências sociales, Barcelona, vol. VII, núm. 146(128), Aug. 2003. Available at:< http://www.ub.es/geocrit/sn/sn-146(128).htm>. Accessed on: 2 July 2009.

CASTELLS, Manuel. The urban question. Translated by Arlene Caetano. Rio de Janeiro: Paz e Terra, 1983. 590p.

CORRÊA, Roberto Lobato. O espaço urbano. 3 ed. São Paulo: Ática, 1995. 94p.

. Constructing the concept of the average city. In: SPOSITO, Maria Encarnação Beltrão. Medium-sized cities: spaces in transition. São Paulo: Expressão Popular, 2007. p. 23-34. (Geography on the move).

DEUS, João Batista de. The Current Structural Transformations in the Economy of Goiás and Their Effects on Socio-Spatial Changes. In: ALMEIDA, M. G. (Org.) Geographical Approaches to Goiás: the natural and the social in contemporary times. Goiânia: IESA, 2002. p. 177-196.

. The Southeast of Goiás: Territorial Transformations Resulting from Brazilian Industrial Deconcentration. 2002. 287 f. Thesis (Doctorate) - Human Geography Course, Usp, São Paulo, 2002a.

ESTEVAM, L. O Tempo da transformação: estrutura e dinâmica da formação económica de Goiás. Goiânia: Ed. do Autor, 1998. 238.

FERREIRA, I. C. B. Ceres and Rio Verde: Two Moments in the Process of Expansion of the Agricultural Frontier. In: AUBERTIN, C. (Org.) Fronteiras. Brasília: Editora Universidade de Brasília; Paris: ORSTOM, 1988. p. 38-59.

GALINDO, O.; SANTOS, V. M. Centre-West: Recent Developments in the Regional Economy. In: AFFONSO, R. B. & SILVA, P. L. B. (Orgs.). Regional Inequalities and Development. São Paulo: FUNDAP/UNESP, 1995. p. 157-194.

GOMES, Paulo César da Costa. Geography and modernity. Rio de Janeiro: Bertrand Brasil, 1996.

GOTTDIENER, Mark. The social production of urban space. 2 ed. São Paulo: Edusp, 1997. 311 p. Translated by Geraldo Gerson de Souza.

GRAZIANO NETO, F. Questão agrária e ecologia: crítica da moderna agricultura. 2. ed. São Paulo: Brasiliense, 1985. 156p.

GUIMARÃES, E. N.; LEME, H. J. C. Caracterização Histórica e Configuração Espacial da Estrutura Produtiva do Centro- Oeste. Campinas: UNICAMP, Population Studies Centre, 1998. p.25-65.

IBGE. 1991 demographic census. Rio de Janeiro: IBGE, 1992.
. 2000 Demographic Census: information base by census sector: Goiás-Jataí-5211909. Rio de Janeiro: IBGE, 2002.

. Cities. Available at:< http://www.ibge.gov.br> Accessed on: 27 Aug. 2007a.

. Regions of influence of cities. Rio de Janeiro: IBGE, 2007b.

. 2000 Demographic Census. Available at:< http://www.ibge.gov.br> Accessed on: 09 June 2009.

. IBGE indicators: national consumer price index system: IPCAINPC. Available at:< http://www.ibge.gov.br> Accessed on: 01 Feb. 2009.
ANTÔNIO HOUAISS INSTITUTE. Houaiss Electronic Dictionary of the Portuguese Language. SI: Objetiva, 2002. CD-ROM.

INTERROGAÇÃO - JEANSWEAR. Company's publicity website. Available at:< http://www.interrogacao.com.br> Accessed on: 12 September 2007.

JATAÍ-GO. Department of Works and Urban Planning. Urban Planning Division.

Jataí: Vacant plots. 1995.

. Department of Works and Urban Planning. Urban Planning Division. Urban Masterplan. 2001.

. Department of Works and Urban Planning. Urban Planning Division. Urban Masterplan. 2003.

. Department of Works and Urban Planning. Urban Planning Division. Cadastral maps. 2003a.

. Department of Works and Urban Planning. Urban Planning Division. Cadastral maps. 2004.

. Superintendence of Industry, Commerce and Tourism. Jataí dossier. 2004a. Available at: < www.jatai.go.gov.br> Accessed on: 26 January 2004.

. Department of Works and Urban Planning. Urban Planning Division. Jataí: Vacant plots. 2004b.

. Fiscal Management Secretariat. Programme for registering economic activities. 2004c. Jataí: [s.n.], 2004.

. Fiscal Management Secretariat. Report on registered economic activities. 2004d. Jataí: [s.n.], 2004.

. Department of Works and Urban Planning. Urban Planning Division. Jataí - Building Standards. 2007.

. Department of Works and Urban Planning. Urban Planning Division. Urban Masterplan. 2007.

. Department of Works and Urban Planning. Urban Planning Division. Amended Urban Masterplan. 2008.

. Department of Promotion and Social Assistance. Bolsa Família Programme. 2008

. Jataí City Hall: photo gallery. Available at:< http://www.jatai.go.gov.br> Accessed on: 24 Apr. 2009.

KAZTMAN, Ruben. Seducidos y Abandonados: el aislamiento social de los pobres urbanos. In: MOYSÉS, Aristides. City, Urban Segregation and Planning. Goiânia: UCG, 2005. p.205-243.

LEFEBVRE, Henri. The right to the city. São Paulo: Moraes, 1991. 145 p. Translation by Rubens Eduardo Frias.
. The city of capital. Translated by Maria Helena Rauta Ramos and Marilena Jamur. Rio de Janeiro: DP&A, 1999. 180p.

LOJKINE, Jean. The capitalist state and the urban question. Translation by Esteia dos Santos Abreu. 2 ed. São Paulo: Martins Fontes, 1997. 359p.

LOUIS DREYFUS COMMODITIES. Company website. Available at:< www.ldcommodities.com> Accessed on: 12 September 2007.

MACHADO, V. F. Southwest Goiás: uneven development. 1996, 143 p. Dissertation (Master's in History) - Federal University of Goiás, Goiânia, 1996.

MARTINE, George. The spatial redistribution of the Brazilian population during the 80s. Brasília: IPEA, 1994. 46p. Available at:< www.ioea.gov.br> Accessed on: 19 April 2004.

MARTINS, J. S. Capitalismo e tradicionalismo: estudos sobre as contradições da sociedade agrária no Brasil. São Paulo: Pioneira, 1975. p. 43-50.

MARTINS, J. S. Fronteira: a degradação do outro nos confins do humano. São Paulo: Hucitec, 1997. p. 145-203.

MARX, K., ENGELS, F. The German Ideology. 7.ed. São Paulo: Hucitec, 1989.

MELO, N. A. Interação campo-cidade: a (re)organização socioespacial de Jataí no período de 1970 a 2000. 2003, 179 p. Master's dissertation - Federal University of Uberlândia, Institute of Geography, Uberlândia, 2003.

MIZIARA, F. Structural conditions and individual choice in the formulation of the concept of "agricultural frontier" In: SILVA, L. S. D. (Org.).

City-country relations: frontier. Goiânia: Cegraf, 2000. p. 273- 289.

MONTE-MÓR, Roberto Luís M. A QUESTÃO URBANA E O PLANEJAMENTO URBANO-REGIONAL NO BRASIL CONTEMPORÂNEO. Belo Horizonte: Cedeplar-Ufmg, 2004. Ministry of National Integration - Secretariat for Regional Development Policies.

MONTEIRO, Carlos Augusto de Figueiredo. The physics of geography: messengers and carriers. Fortaleza-CE: Multigraf, 1995. p. 5-57.

OLIVEIRA, José Aldemir de. The city on the horizon of the probable: policies and urban development. In: SILVA, José Borzacchiello da; COSTA, Maria Clélia Lustosa; DANTAS, Eustógio Wanderley C. (Orgs). The city and the urban. Fortaleza: EUFC, 1997. p.265-271

PALACÍN, L. Goiás 1722/1822: structure and conjuncture in a mining captaincy. Goiânia: Gráfica Oriente, 1972. 177 p.

; MORAES, M. A. S. História de Goiás. 5. ed. Goiânia: UFG, 1989. 125 p.

; GARCIA, L. F.; AMADO, J. History of Goiás in documents: I. Colony. Goiânia: UFG, 1995. 222 p. (Documentos Goianos, 29).

PELUSO, Marília Luiza. Home ownership and the dream. In: SILVA, José Borzacchiello da; COSTA, Maria Clélia Lustosa; DANTAS, Eustógio Wanderley C. (Orgs). The city and the urban. Fortaleza: EUFC, 1997. p.235-243.

PNUD/IPEA/FUNDAÇÃO JOÃO PINHEIRO. Atlas of human development in Brazil. Version 1.0.0: ESM Consultoria, 2003.

RIBEIRO, Luiz Cesar de Queiroz. Residential segregation: theories, concepts and techniques. In: MOYSÉS, Aristides. City, Urban Segregation and Planning. Goiânia: UCG, 2005. p.89-126.

RODRIGUES, Juciano Martins. Population, Society and Territory: Time of Urbanisation in Goiás. 2006, 132p. Dissertation (Master's in Population Studies and Social Research) - ENCE/IBGE, Rio de Janeiro, 2006.

SANTOS, José Alcides Figueiredo. Social Class and Gender Inequality in Brazil. Dados: Revista de Ciências Sociais, Rio de Janeiro, v. 51, n. 2, p.353-402, 2008.

SANTOS, Milton. Brazilian Urbanisation. São Paulo: Editora Hucitec, 1993.

. The nature of space: technique and time, reason and emotion. 3. ed. São Paulo: Hucitec, 1999. 308 p.

. The citizen's space. 5.ed. São Paulo: Studio Nobel, 2000. 142p.

. SILVEIRA, Maria Laura. O Brasil: território e sociedade no início do século XXL 3 ed. Rio de Janeiro: Record, 2001.473p.

SAUER, C. O. The Morphology of the Landscape. In: CORRÊA, R. L.;
ROSENDAHL, Z. Paisagem, Tempo e Cultura. Rio de Janeiro: EdUERJ, 1998. p. 12-74.

SCOPEL, Iraci et al. Evolution of land use in the Córrego do Açude watershed, in Jataí-GO. Boletim Goiano de Geografia - Institute of Socio-Environmental Studies - Research and Postgraduate Programme in Geography - Geography Course - UFG. Goiânia: Editora da UFG. v.22(2): 31-46. juL/dez. 2002.

SEABRA, Odette Carvalho de Lima. Urbanisation and fragmentation: notes for the study of the neighbourhood and urban memory. In: SPOSITO, Maria Encarnação Beltrão. Urbanisation and cities: geographical perspectives. Presidente Prudente: [s. n.], 2001. p. 75-95.
SEPLAN. Conjuntura Económica Goiana, n. 02 - (2004). Goiânia: Planning and Development Secretariat, 2004.

. Ranking of Goiás Municipalities: 2005. Goiânia: SEPLAN, 2005.

SILVA, José Borzacchiello da. Discussing the city and the urban. In: SILVA, José Borzacchiello da; COSTA, Maria Clélia Lustosa; DANTAS, Eustógio Wanderley C. (Orgs). The city and the urban. Fortaleza: EUFC, 1997. p.265-271

. City statute versus city statute: that is the question. In: CARLOS, Ana Fani Alessandri; LEMOS, Amália Inês Geraiges (eds.) Dilemas Urbanos: novas abordagens sobre a cidade. São Paulo: Contexto, 2003. p.29-34.

SILVA, Márcio Rodrigues. Encounters and mismatches: a study of the urban space of Jataí-GO. 2005, 113p. Dissertation (Master's in Geography) - Federal University of Goiás, Institute of Socio-Environmental Studies, Goiânia, 2005.

et al. History of occupation and current land use in the micro-basin of the córrego do sapo in Jataí-GO. Revista do Departamento de Geografia da UFMT, Rondonópolis, v.2(2): 117-125. 2002.

SILVA, Sinara Rosa Carvalho e. Cicatrizes urbanas: marcas de socialização e formação na cidade em Jataí-GO. 2009, 136f. Dissertation (Master's in Education) Federal University of Goiás, Faculty of Education, Goiânia, 2009.

SOARES, Beatriz Ribeiro. Small and medium-sized towns: A study of socio-spatial relations in the cerrado areas of Minas Gerais. In: SPOSITO, Maria Encarnação Beltrão. Medium-sized cities: spaces in transition. São Paulo: Expressão Popular, 2007. p. 461-494. (Geografia em movimento).

SOUZA, Marcelo Lopes de. ABC of urban development. Rio de Janeiro: Bertrand Brasil, 2003. 190p.

. Changing the city: a critique of urban planning and management. Rio de Janeiro: Bertrand Brasil, 2003a. 556p.

SOUZA, Maria Adélia A. de Souza. Geography, verticalisation and land valuation. In. Identity of the metropolis: verticalisation in São Paulo. São Paulo: Hucitec/Edusp, 1994. p. 151-173.

STEINBERGER, Marília. The significance of the Centre-West region in the spatiality of Brazilian development: a geopolitical analysis. In: GONÇALVES, Maria Flora;

BRANDÃO, Carlos A.; GALVÃO, Antonio C. (Orgs.). Regions and cities, cities in regions: the regional urban challenge. São Paulo: UNESP-ANPUR, 2003. p.609-620.

TEIXEIRA NETO, A.; GOMES, H. Diagnóstico sócio-econômico do Sudoeste Goiano. Jataí: Prefeitura Municipal de Jataí/Fundação Educacional de Jataí, 1995.

TEIXEIRA NETO, Antônio. The territory of Goiás: formation and process of settlement and urbanisation. In: ALMEIDA, Maria Geralda de (Org.). Geographical approaches to Goiás: the natural and the social in contemporary times. Goiânia: IESA, 2002. p.11-45.

UNESCO. Intermediate Cities and Global Urbanisation. Lleida, Spain: Lleida City Council, UNESCO, UIA, Ministry of Foreign Affairs, 1999.

VILLAÇA, Flávio. Intra-Urban Space in Brazil. 2ed. São Paulo: Studio Nobel: FAPESP: Lincoln Institute, 2001. 373p.

ANNEXES
ANNEX - A

TOURIST MAP OF JATAÍ

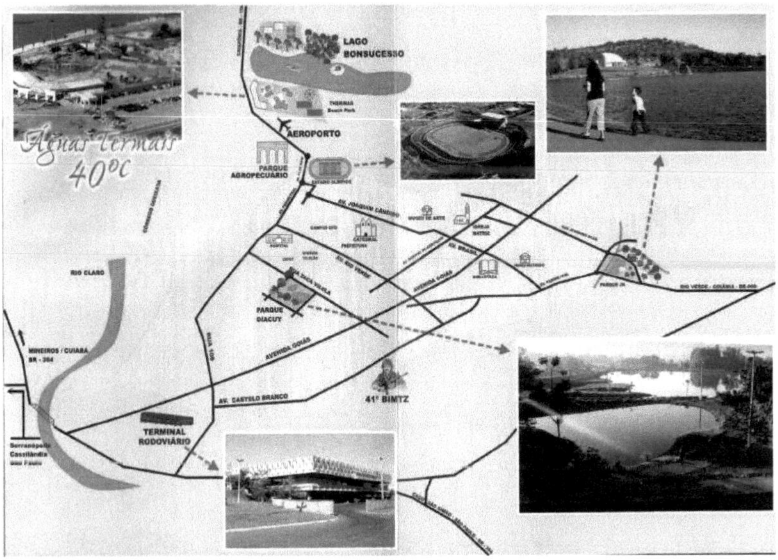

ANNEX - B

PUBLICITY MATERIAL FOR THE PORTAL DO SOL ALLOTMENT

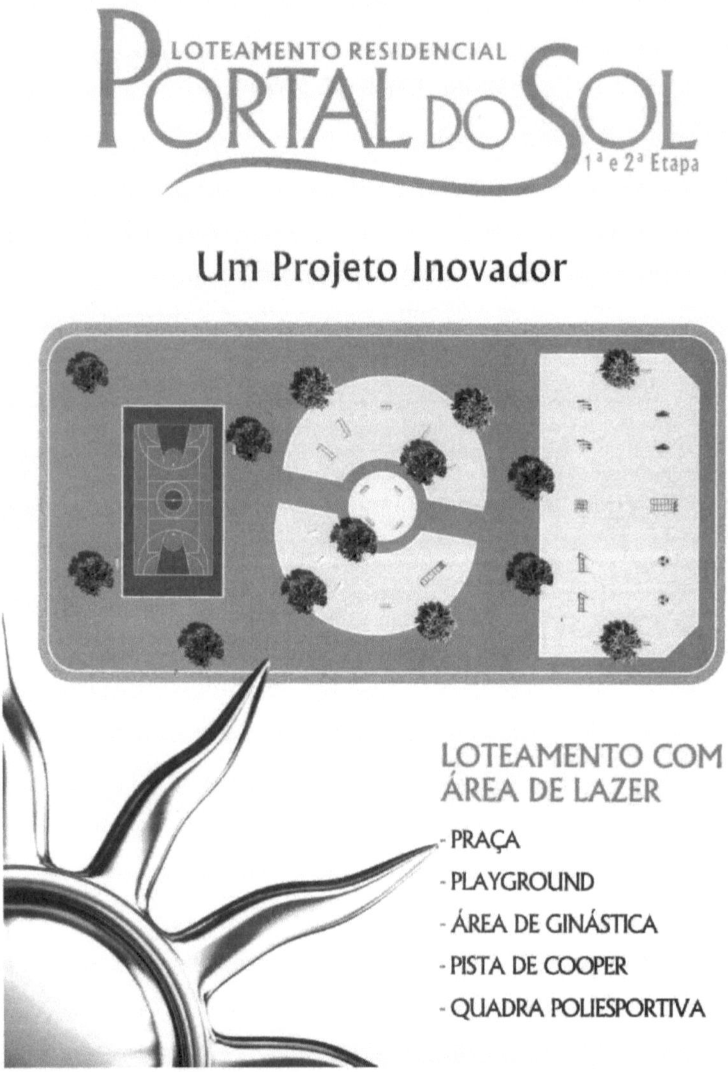

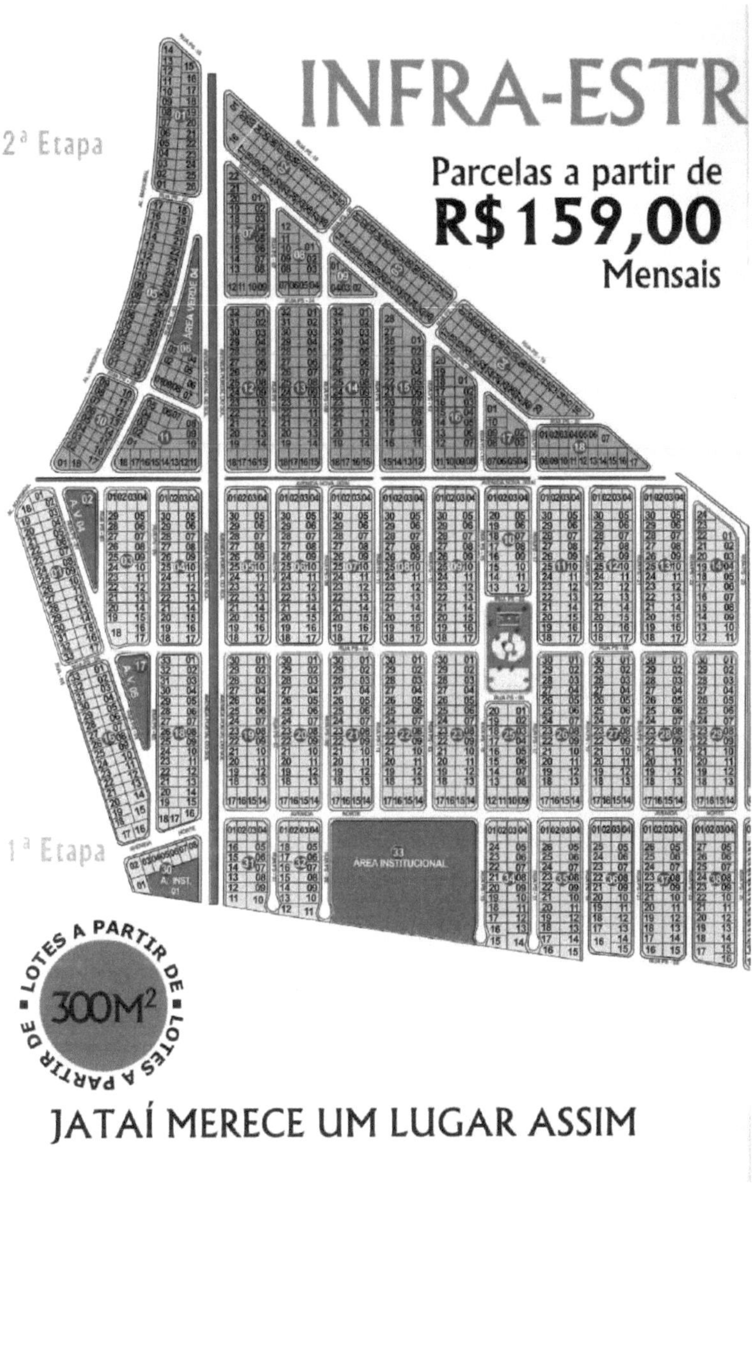
INFRA-ESTR
Parcelas a partir de
R$159,00
Mensais
2ª Etapa
1ª Etapa
ÁREA INSTITUCIONAL
LOTES A PARTIR DE
300M²
LOTES A PARTIR DE
JATAÍ MERECE UM LUGAR ASSIM

UTURA INVEJÁVEL
Agora,você poderá morar com a infra-estrutura e qualidade de vida que sempre sonhou!
- RUAS ASFALTADAS
- MEIO FIO
- ENERGIA ELÉTRICA
- ILUMINAÇÃO PÚBLICA
- ÁGUA POTÁVEL
- LOTES DEMARCADOS
OBRAS EM RÍTMO ACELERADO
MAIS UM LANÇAMENTO
DE SUCESSO
LÍRIOS DO CAMPO
EMPREENDIMENTOS IMOBILIÁRIOS
IMOBILIÁRIA
REI
3636-4400
TORRE - TELEGOIÁS
ÁREA VERDE 12
ÁREA INSTITUCIONAL 03
PLANTÃO DE VENDAS NO LOCAL
Número do Registro Portal do Sol - 1ª Etapa R 03.M / 33988 - 2ª Etapa R 03.M / 34239

Crescer Crescer, Prá Jataí Crescer

ANNEX - C

PUBLICITY MATERIAL FOR THE RESIDENTIAL SUBDIVISION OF BRISAS

Localização sem igual
Ao lado do Residencial Barcelona
CENTRO
RESIDENCIAL BARCELONA
RESIDENCIAL DAS
Brisas

ANNEX - D

PUBLICITY MATERIAL FOR THE MORADA DO SOL HOUSING ESTATE

Adquira seu lote e
construa sua casa em
um lugar alto, arejado,
de excelente localização,
ao lado do TRE e
Condomínio Barcelona.
Investimento
de grande valorização!

MIX
Papier aus verantwortungsvollen Quellen
Paper from responsible sources
FSC® C105338

Printed by Books on Demand GmbH, Norderstedt / Germany